The St. Lawrence Seaway and Power Project

An Oral History of the Greatest Construction Show on Earth

Claire Puccia Parham

SYRACUSE UNIVERSITY PRESS

Syracuse University Press, Syracuse, New York 13244-5160

Copyright © 2009 by Claire Puccia Parham

All Rights Reserved

First Edition 2009

09 10 11 12 13 14   6 5 4 3 2

The paper used in this publication meets the minimum requirements
of American National Standard for Information Sciences—Permanence
of Paper for Printed Library Materials, ANSI Z39.48–1984.∞™

For a listing of books published and distributed by Syracuse University Press,
visit our Web site at SyracuseUniversityPress.syr.edu.

ISBN-13: 978-0-8156-0913-1

Library of Congress Cataloging-in-Publication Data

Parham, Claire Puccia.
The St. Lawrence Seaway and Power Project : an oral history of the greatest
construction show on earth / Claire Puccia Parham. — 1st ed.
p. cm.
Includes bibliographical references and index.
ISBN 978-0-8156-0913-1 (hardcover : alk. paper)
1. Saint Lawrence Seaway—History.   2. Dams—Saint Lawrence River—
Design and construction—History.   3. Hydroelectric power plants—Saint
Lawrence River—Design and construction—History.   4. Waterways—Saint
Lawrence River Region—History.   5. Saint Lawrence Seaway—Power
utilization—History.   6. Construction industry—Employees—Interviews.
I. Title.   II. Title: Saint Lawrence Seaway and Power Project.
TC427.S3P37 2009
627'.13709714—dc22
2008047294

Manufactured in the United States of America

Contents

Illustrations

Acknowledgments

I WOULD LIKE TO THANK all of those individuals who made this book possible. To my husband, Edward, and my daughters, Eve and Annabelle, who patiently supported my long work hours and respected my dedication to this project. To the many men and women who shared their memories with me, without whom this manuscript would not have been possible. To my mother, Juliette Wager Puccia, who encouraged me to recognize what an important time the 1950s were for women. Finally, to my father, Lawrence Puccia, Jr., the hardest-working man I know, who inspired my interest in construction and the plight of the workers who struggle every day to keep our country running.

CLAIRE PUCCIA PARHAM is a native of Watertown, New York, and is currently an instructor in the History Department at Siena College in Albany, New York, specializing in American history and twentieth-century world history. Her first book, *From Great Wilderness to Seaway Towns: A Comparative History of Cornwall, Ontario, and Massena, New York, 1784–2001,* was published in 2004. In it Dr. Parham describes the struggles of early settlers and analyzes the development of industry and the construction of the St. Lawrence Seaway. The comprehensive study also highlights the unique lives of Cornwall and Massena residents, discusses the distinctive society and culture they established, and concludes with a comparison of the long-term economic impact of the Seaway on the region. Dr. Parham received her Ph.D. in American history from Binghamton University, State University of New York; her M.S. in labor studies from University of Massachusetts at Amherst; and her B.A. in government from St. Lawrence University.

Interviewees

Sam Agati founded Laborer's Local 322. He was a native of Syracuse, New York.

Ambrose Andre was hired by the U.S. Army Corps of Engineers upon graduation from Clarkson University to assist in designing the Snell and Eisenhower Locks in the Buffalo District Office. In 1955 he was transferred to the Corps field office in Massena and was later promoted to the position of concrete inspector.

Eunice Barkley worked on her mother's lunch cart outside the gates of the Long Sault Dam. In the evening she worked as a waitress at the Village Inn in Waddington. She married Ron Barkley in 1958.

Ron Barkley was a truck driver for V. A. Roberts, the project manager for Morrison-Knudsen, at the Red Mills section of the Galop South Channel improvement. He was also employed as a painter on the Iroquois and Long Sault Dam projects.

Irene Bryant was married to John Bryant. She was trained as a veterinarian, but stayed at home to raise her children during the Seaway project.

John (Jack) Bryant had worked for the Bureau of Reclamation on the Hungry Horse Dam and handled construction claims in the chief engineer's office in Denver. He was hired along with many of his former co-workers from the Bureau to manage the claims adjustment department for Uhl, Hall, and Rich.

Helen Buirgy was the wife of Ralph Buirgy, who served as an office engineer for Uhl, Hall, and Rich. She was born and raised in Nebraska on a farm and ranch, and taught school in a one-room schoolhouse before her marriage.

Robert Carpenter, a native of Carthage, New York, graduated from Clarkson University and became an engineer for the Corps of Engineers. He initially assisted with preliminary subsurface investigation both in the Buffalo District Office and on site in Massena. During the Seaway construction, Carpenter served in the Massena District Office as one of five area engineers.

Theodore (Ted) Catanzarite graduated from Massena High School in 1955 and worked on the Seaway for two years as a laborer until he entered university in September 1957.

James (Jim) Cotter had worked for the Bureau of Reclamation in the western United States, on the Davis Dam on the Colorado River and the Hungry Horse Dam in western Montana. The Uhl, Hall, and Rich employment office hired him to deal with contract administration from 1955 to 1959. He had also previously worked as an insurance claims adjuster.

Phyllis Cotter was married to Jim Cotter. She was trained as a medical technician, but took a break from her career to raise her children during the Seaway project.

Joseph Couture worked as a carpenter on numerous projects for Ontario Hydro before relocating to Cornwall. He was transferred to the power dam project to work with Iroquois Constructors, the main contractor on the site. Couture initially worked as a carpenter, but was later promoted to the rank of foreman on the downstream end of the dam.

Les Cruickshank spent ten years as an equipment operator for Ontario Hydro, first at the Niagara Falls project and then on the Seaway Project. He became an independent contractor and purchased his own grader to level land at the township relocation areas known as the Lost Villages.

Glenn Dafoe was employed by Ontario Hydro as a carpenter on the houses relocated to the new town sites. He had previous carpentry experience and continued to work in his trade after the completion of the Seaway.

Shirley Davis relocated to Waddington, New York, with her husband, Dick, who initially found employment at Alcoa. He was hired as an engineer on the Seaway project and the couple remained in the area until the project's completion.

John Dumas was the son of well-known local reporter Eleanor Dumas, who worked for the *Watertown Daily Times*. He is a lifelong resident of northern New York.

Cyril Dumond served as a pipe fitter on the Robert Saunders Power Dam. He remained on the project for its entirety and continued his career in Cornwall after the project's completion.

Joyce Eastin moved to Massena on February 22, 1955, with her husband, Howard, who was the project manager at the Iroquois Dam in Waddington, New York, and their children. They remained for three and a half years.

Lowell Fitzsimmons was a dragline operator and member of Operating Engineers Local 545. He worked on the project from 1955 to 1958.

J. David (David) Flewelling worked as a draftsman for B. Perini, the managing contractor on the Snell Lock doing rebar lift drawing. After his layoff from that position, he spent his last few months in Massena laboring on the Barnhart Island project.

Joseph Foley served as an engineer in the Buffalo Corps of Engineers District Office. In this capacity he revised the concrete drawings for the Eisenhower and Snell Locks. He continued his career with the Corps for several decades, focusing on the annual rehabilitation of the Snell and Eisenhower Locks until his retirement in 1983.

William (Bill) Goodrich was the human resources director for Ontario Hydro, a position he had previously held on the Niagara Falls project. Besides overseeing employment issues, he created numerous sporting activities for Hydro employees.

Floyd Grant served as a security guard on the various construction sites on the American side. He still resides in Massena, New York.

George Haineault worked as a carpenter for Ontario Hydro on the power dam project until its completion. Previously he was employed by contractors on local construction projects as a second-year apprentice.

Kenneth Hallock began his career with the Corps of Engineers in the Buffalo District Office as a junior engineer. He participated in the redrawing of the original Seaway plans completed in 1941. During the project, he was appointed chief structural engineer for all of the Corps construction sites on the American side of the project, which included the Eisenhower and Snell Locks.

Barbara Hampton was married to Robert Hampton. She was originally from Albany, New York.

Robert Hampton conducted hydrographic surveys on the St. Lawrence River for the Corps of Engineers in 1954. His past work experience included surveying various road and utility projects in New York State. When construction commenced on the Seaway, he served as the contract supervisor of the upstream channel enlargement and dredging contracts.

Keith Henry was an eight-year veteran of the Ontario Hydro Hydraulic Department by the commencement of the St. Lawrence Seaway and Power Project. He conducted water level tests for the Ottawa River and the Niagara Falls projects, and supervised the building of the hydraulic models for the Seaway. He was transferred to Cornwall in 1955, where he served as the river control engineer. Henry continued his career with

Ontario Hydro after the completion of the St. Lawrence project and served as a commissioner on the International Joint Commission from 1972 to 1978.

Dolores Kormanyos worked as a waitress at the Cornwallis Hotel during the luncheon for the dedication ceremony attendees. She had emigrated to Massena in 1953 after her marriage.

David Manley began working as a laborer on the Snell Lock in 1956 when he was a junior at Malone High School. He also helped construct the vehicle tunnel that runs under the Eisenhower Lock.

Ann Marmo was married to Joseph Marmo. She was trained as a nurse, but was a homemaker while in Massena.

Joseph (Joe) Marmo spent seven years in Massena working as an office manager for the Power Authority and as a contractor claims evaluator for Uhl, Hall, and Rich. Before his employment on the project, he had received a degree in dam design from the University of Michigan.

Cornelius (Neil) McKenna worked as a carpenter on the Long Sault Dam, the Snell Lock, and the Eisenhower Lock between November 1956 and December 1957.

Alfred Mellett held a civil engineering degree from the Massachusetts Institute of Technology and Lowell Institute. Before the St. Lawrence Seaway and Power Project, he had worked on several other large hydrodam projects including one in Turkey. While in Massena, he initially conducted surveys for Uhl, Hall, and Rich and eventually served as the official photographer for the New York State Power Authority.

Ira Miller worked as a carpenter and garbage barge operator at the main power dam and Long Sault Dam site. He was a native of Nicholville, New York, and had previously worked as a truck driver.

Hubert Miron was the head payroll clerk for Ontario Hydro. He supervised the collection of weekly payroll sheets and delivered paychecks to all of the agencies' workers on the Canadian construction sites every Friday.

Garry Moore, a native of Cornwall, Ontario, held various positions with the St. Lawrence Seaway Authority for four years beginning in the summer of 1955, including boat operator, hydrographic survey crew member, and water-level monitor. In his last few months on the project, he was assigned to manage the surveying for the dredging of the Seaway channel in Lake St. Francis.

John Moss was hired as a concrete inspector for Ontario Hydro with no previous experience. He had formerly been employed at Domtar, the local paper mill in Cornwall.

Arthur Murphy began his career with Ontario Hydro when he was in high school. He worked part-time in the summer in the warehouse logging in sections of cranes and machinery. Upon graduation, he was hired as a tour guide for the various construction sites by the directors of Ontario Hydro's information section.

Roderick Nicklaw was a student at Malone High School when he began working as a laborer at various construction sites on the American side.

Donald Rankin served as divisional engineer at the power house project for Iroquois Constructors, the supervising contractor for Ontario Hydro. He was second in command to the project engineer. Rankin had previously worked as a miner and as a construction supervisor for Ontario Hydro.

Frank Reynolds served as a dredging supervisor for Morrison-Knudsen. His prior work experience included performing computer and surveying work for the Department of Defense.

Thomas Rink, Ann Marmo's brother, gained entry into the laborers' union based on the recommendation of his brother-in-law, Joe. He remained in Massena until 1958 and was employed by various contractors on the American construction sites.

James Romano owned A. J. Trucking. He and the drivers under his command transported gravel to the various construction sites from off-site pits. Romano only worked on the Seaway during the winter and spent the remainder of the year operating his trucking business in his hometown of Rochester, New York.

William Rutley worked as an engineer's assistant at the various town sites constructed by Ontario Hydro. He measured and assisted in the surveying of homes before their relocation to new lots and placement on new foundations.

J. Arnold Shane initially worked as a truck driver for Iroquois Constructors on the Robert Saunders Power Dam while still a high school student. Upon graduation, he worked in Ontario Hydro's stationary shop and was promoted to the position of junior concrete inspector.

Thomas Sherry was hired by B. Perini in 1956 through Masons Local 81 to work on the Eisenhower Lock. He later faced with sandstone the power dam administration and tourist building and the Massena town beach clubhouse.

Roy Simonds assumed the role of chief of contracts for Uhl, Hall, and Rich, the consulting engineers for PASNY for 1956–60. In this capacity, he along with his co-workers administered contracts and oversaw contractor claims adjustments. He had previously worked for the Bureau of Reclamation.

Melba Singleton married Ray Singleton in 1956. Before her marriage she worked as a chief operator for Bell Telephone in Cornwall, and during the project she worked in the family's restaurant.

Ray Singleton worked on the first bridge over the Grasse River, the Long Sault Dam, and the main power dam as a gantry crane and heavy equipment operator. Eventually his father and three brothers came to Massena to operate heavy equipment on the various project sites.

William (Bill) Spriggs served as an electrical inspector for the Corps of Engineers, initially setting up the electrical supply lines for contractors. For the remainder of the project he inspected the installation of power sources at the Snell and the Eisenhower Locks.

Barbara Brennan Taylor was born and raised in Cornwall. Between 1955 and 1958 she served as the secretary for the office manager for Iroquois Constructors, the main contractor for the Canadian side of the power dam. Previously she was a secretary in the administrative offices of Courtaulds, the local division of a British textile producer.

Bart Whitten served as a geological supervisor for Ontario Hydro on the power dam project. He came to Cornwall from the Niagara Falls project, where he served as a carpenter.

Frank Wicks grew up in Canton, New York, and began working as a laborer in 1957. He was eventually employed by Uhl, Hall, and Rich as a member of a concrete crew.

Marjorie (Marge) Wiles was married to Don Wiles, an engineer for Uhl, Hall, and Rich. She was a native of Omaha, Nebraska, and lived in the Buckeye Development with her husband and children.

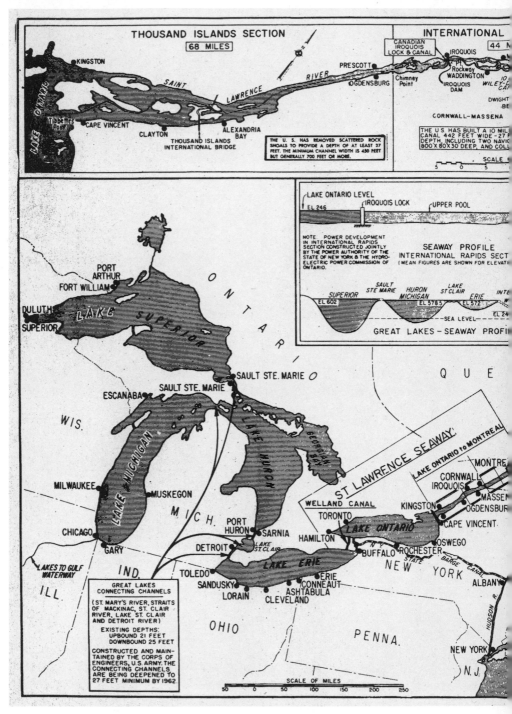

1. Map of the Great Lakes–St. Lawrence Seaway project.
Courtesy of St. Lawrence Seaway Development Corporation.

Introduction

AWESOME! BREATHTAKING! A feat of modern engineering! These are some of the words people exclaim as they overlook the Robert Moses–Robert H. Saunders Dam, which straddles the U.S.–Canadian border. The 3,216-foot-long, 195.5-foot-high structure was the focal point of the billion-dollar St. Lawrence Seaway and Power Project, the largest waterway and hydrodam ever constructed jointly by two nations. Planners had scheduled the construction of the St. Lawrence Seaway and power dam to take seven years; however, ship captains navigated the new waterway and the turbines rotated only four years after Robert Moses, chairman of the Power Authority of the State of New York (PASNY), had dug the first chrome spadeful of dirt at the 1954 ground-breaking ceremony. The project was the most important public works initiative of the Eisenhower administration and the Cold War. The completed inland waterway protected ships and submarines in the event of an attack, and the power dam provided a new source of electricity for American and Canadian residents and manufacturers.

M. W. Oettershagen, deputy administrator of the Seaway Development Corporation in 1959, dubbed the St. Lawrence Seaway and Power Project "the greatest construction show on earth." Consisting of seven locks, the widening of various canals, and the taming of rapids, the 265-mile Seaway project between Alexandria Bay, New York, and Montreal, Canada, fulfilled politicians' and engineers' century-long dreams of linking the Great Lakes interior industrial hubs in the United States to the Atlantic Ocean. Rivaling the great dam and waterway projects of the 1920s, the Seaway introduced new state-of-the-art equipment and required the participation of the seven largest U.S. contractors and 22,000 carpenters, engineers, and laborers. Ontario Hydro and the U.S. Army Corps of Engineers (Corps) employees toiled for decades conducting soil testing, building models, and designing locks and dams. In his May 1959 report entitled *St. Lawrence Seaway—Fact and Future*, Oettershagen concluded that during the construction between 1954 and 1958, there had been a "deployment of an estimated $75 million in on-site equipment, the placing of more than six million cubic yards of concrete and the dredging and excavation of 360 million tons of materials."[1] William Willoughby, author of *The St. Lawrence Waterway: A Study in Politics and Diplomacy*, asserted that the waterway and hydroelectric facility project was one of the most incredible engineering and construction jobs ever attempted.[2]

The construction of the numerous project sites presented the employees of Uhl, Hall, and Rich, the project managers for PASNY, and the Corps, the agency in charge of completing the locks and dredging on the American side of the project, with many logistical dilemmas. The short lag time between the proposal of the Wiley-Dondero Act in 1951 and the commencement of excavation in 1954 prevented designers from preparing final detailed drawings and scientists from completing comprehensive test borings. Therefore, for the next four years, these entities and their contractors faced a record number of change orders and design modifications, and encountered clay, glacial till, and rocks on all of the construction and dredging sites that destroyed machinery and bankrupted excavation contractors. Contractors also did not anticipate how the extreme weather conditions would adversely impact workers, equipment, and concrete. During the dredging of the power pool and new navigational route, Ontario Hydro and Corps river engineers additionally had to maintain a consistent water level to prevent cofferdams from washing away and to allow uninterrupted use of the existing shipping channels.

Eleanor Dumas, a newspaper reporter during the construction, recollected, "'It was the best of times, it was the worst of times.' . . . Looking back over thirty years since the groundbreaking for what was then the world's largest construction project, the best of times came for the hard workers, the money makers and the achievers. . . . The worst to those people on each side of the St. Lawrence River who mourned the loss of their old cemeteries and the Long Sault Rapids. . . . The initial work started unobtrusively." However, she then noted, as the excavators cleared lock and dam sites, "the earth turned into a terror scene" as engineers encountered unfamiliar substances that "machines found hard to penetrate."[3]

In a June 26, 1958, article in the *Chicago Daily News,* a reporter compared the St. Lawrence Seaway dimensions and construction, excluding the hydroelectric project, with the forty-four-year-old Panama Canal. In terms of the size of the actual construction area, the Seaway was 114 miles long, compared to the 50.5-mile-long Canal. The Seaway locks lifted ships 580 feet versus 85 feet for the Canal. More than one hundred million dollars differentiated the construction cost of the two projects, but that can be explained by the four-decade rise in labor costs and materials. A workforce of 39,000 completed the Canal in eight years, double the time it took 22,000 men to construct the Seaway. However, oceangoing vessels could traverse the Panama Canal because of its five additional 400-foot by 110-foot locks. Additionally, the Canal's 45-foot channel depth exceeded the Seaway's 27-foot depth. Even though they differed in terms of cost and size, both waterways had long been recognized by world leaders as vital to North American commerce. Regardless, both faced many false starts and hurdles in terms of gaining political approval and financial backing.

Following the completion of the Panama Canal in 1914, Bertrand Snell, a member of the U.S. House of Representatives, proposed the first piece of St. Lawrence Seaway legislation, but for five decades, members of the Canadian Parliament and congressmen hotly debated the expense and merits of the construction of the waterway and power dam. According to Lionel Chevrier, a former Canadian cabinet member and head of the St. Lawrence Seaway Authority from 1954 to 1957, "Few projects have been so bitterly opposed or inspired so many opinions, argument, legal battles, treaties, and inter-government memoranda."[4] During the early twentieth century, Canadian national and provincial political leaders—specifically Maurice Duplessis, the premier of Montreal—did not see the urgency of building a new seaway and dam when the nation's existing canal system and power facilities fulfilled transportation and electrical demand. American presidents since the administration of Woodrow Wilson supported the Seaway construction, but lobbyists for the railroads and Atlantic ports and entrepreneurs criticized the project because they feared the completed waterway would result in the loss of business or other injury. Opposition on both sides of the border along with the involvement of American and Canadian forces in both world wars and the economic downturn during the Great Depression delayed the funding and the approval of the final project plans for several decades. During the 1930s and early 1940s, President Franklin Roosevelt and Canadian Prime Minister Mackenzie King focused instead on war mobilization efforts and on unemployment and social relief programs.

A shift in America's role on the world stage precipitated the passage of the Wiley-Dondero Bill by the American House of Representatives and Senate in 1951 and 1954 respectively, which approved the final plans and funding of the American section of the St. Lawrence Seaway and Power Project. While supporters of the project had always touted it as a means to improve domestic transportation routes and power production, federal political leaders now saw it as a key element in their crusade against communism. The commencement of construction in 1954 dovetailed with a shift in American defensive strategy and that of its Communist foes. The previous year President Harry Truman had declared an armistice with Korean leader Kim Il Song, effectively ending the three-year Korean conflict in which the U.S. military faced its first defeat in its history based on inadequately trained troops and archaic weaponry. Joseph Stalin also declared that he now possessed the hydrogen bomb. While Stalin had improved Russia's industrial might and navigational facilities with his Five-Year Plans in the 1920s and 1930s, President Franklin Roosevelt had delayed such improvements and instead focused his New Deal programs on relieving social strife. Truman and his successors needed to expand U.S. industrial capacity to satisfy new consumer demands and to rebuild and sustain a prosperous Europe in order to prevent the further spread of communism. The St. Lawrence Seaway and Power Project was an important element in the transition of the United States and Canada from isolationist nations concerned with domestic economic stability to post–World War II democratic superpowers focused on global military and industrial dominance.

Cold War politics and the American-Soviet rivalry explain the eventual passage of the Seaway legislation; however, they did not impact the day-to-day lives of the workers and their families. Instead, the unprecedented cooperation between Canadian and American agencies and contractors made the completion of the St. Lawrence Seaway and power dam possible. The creation of a unified and productive workforce by Seaway contractors from a group of carpenters, laborers, and engineers with varying cultural and economic backgrounds illustrated a unique accomplishment. While experience levels between American and Canadian workers differed, their strong work ethic crossed the border. When these men stepped onto the job site, they put aside their personal issues and strove to complete what they saw as the most important job of their lives. The cultural clashes between workers could have resulted in more accidents and defects in workmanship. However, these actions would have not only impacted the quality of the construction, but a worker's employment status. Union leaders maintained a strong hold over the behavior of their members employed on the project sites on both sides of the border and addressed periodic worker shortages and disagreements with contractors in a timely fashion to limit the number of strikes. Laborers and carpenters who caused conflicts on the job or were involved in an off-site fight could be fired or blacklisted for the remainder of the project's construction.

The different religious, economic, and social backgrounds of workers and their families at first created tension between themselves and native Massenans. Many workers arrived from urban areas and had become accustomed to modern conveniences and to accepting moral and social attitudes. They encountered a population in Massena and Cornwall who cherished small-town values and traditional forms of religious worship. Once local residents realized the nonthreatening nature of these temporary members of their community, religious congregations accommodated new members and welcomed the establishment of new faiths. To satisfy the more sophisticated tastes of Seaway engineers and their wives, store owners and restaurant and bar proprietors expanded their selections to include more expensive products and earned a hefty profit. At the end of the four-year project, workers had left their impression on a community that had been culturally and economically stagnant for several decades.

My personal interest in the construction of the St. Lawrence Seaway and Power Project began twenty years ago. In 1989 I enrolled in an undergraduate research seminar at St. Lawrence University. The professor, Dr. Alan Draper, charged the students with exploring a new archival collection concerning the St. Lawrence Seaway and Power Project. I concentrated on those documents that addressed the lives of the workers and the social impact of the project on Massena, New York. As part of this assignment, Dr. Draper supplied me with the names of ten workers who had volunteered to discuss their memories of working on the project. I incorporated those conversations into a brief paper that compared how the men's recollections varied from newspaper accounts and official accident statistics. At the end of the semester I, along with my fellow students, presented our findings to a public gathering hosted by the St. Lawrence County Historical Society.

When I entered the doctoral program at Binghamton University in 1995, I assumed I had an excellent dissertation topic. However, over the next several years I learned that although the construction of the Seaway and the experience of local residents and workers interested me, the topic did not meet the stringent requirements for a dissertation topic because of its narrative nature and its lack of an analytical framework and argument. I therefore wrote a comparative history of Cornwall, Ontario, and Massena, New York, entitled *From Great Wilderness to Seaway Towns,* which was published in 2004.

After I completed my graduate studies, the tales of the gruesome accidents, bone-chilling weather, and larger-than-life equipment used on the construction haunted me. These men remembered their day-to-day routine and what they had eaten for lunch fifty years earlier. For many this would be their first and only job on a major public works project. Based on their mutual participation in this life-altering experience, these men established enduring friendships.

When I decided to continue my oral history project in 2004, I ran into a major problem. None of the contractors had saved their employment records from the 1950s, and neither had Ontario Hydro, PASNY, or the Seaway Authorities. Newspaper articles published in the *Massena Observer* and the *Cornwall Standard* mentioned my ongoing research and my desire to find workers interested in being interviewed. I received numerous telephone calls and e-mails from former workers, as well as from their children who had been raised on stories about the construction and significance of the St. Lawrence Seaway and Power Project. I interviewed fifty-three American and Canadian workers and their wives, some of whom are now deceased.

These oral interviews, along with newspaper articles and monthly construction reports, comprise the bulk of my sources. I obtained secondary sources from the St. Lawrence University Archives, the Bertrand Snell Collection at the State University of New York at Potsdam, and private collectors. I taped each interview and obtained permission from each subject to use his or her comments. I had a list of specific questions, but most of the time the free-flowing conversation garnered the answers and additional information without any prompting. Interviewees were sincere and excited that I was interested in their stories. Each person I interviewed also provided me with the contact information for at least two other people. My first interviewee, Bart Whitten, a concrete inspector on the Robert H. Saunders Power Dam, compiled a detailed list of fifteen other workers, as well as their addresses, phone numbers, and job titles on the Canadian side of the power project.

I realized that the use of oral sources, however, was not without its pitfalls of which I, as the interviewer, needed to be cognizant. Many of the workers' and their wives' memories may have become clouded or exaggerated, given that the construction of the Seaway ended fifty years ago. Memory is subjective, and therefore cannot always be interpreted as absolute. I have tried, where possible, to verify the interviewees' allegations in official documents and statistics. Based on the fact that many of my interviewees worked together and still communicated with each other, I expected to find redundancies in their responses, but have used them to emphasize a theme or point.

I am aware of the problems presented by my limited sample of interviewees. However, this situation is based on several variables and did not affect the overall validity of my study. First of all, many of the individuals who held management positions are deceased, given that the project was completed in 1958. Others, including Native American ironworkers, were elusive. Those included in this study were willing participants who contacted me or to whom I was referred by an acquaintance. My pool does, however, provide accounts of various levels of workers from laborers to the divisional engineer on the Robert H. Saunders Power Dam. The inclusion of women also adds diversity.

The transcription of these interviews offered an inside view of the men under the bright yellow hard hats, who worked day in and day out, in all weather conditions, to complete the largest waterway and power dam project in North America. Why did it take more than fifty years of political debate and engineering studies to get the project built? What were the major structures constructed during the St. Lawrence Seaway and Power Project? Who were the laborers, carpenters, and engineers who completed the locks, dam, and dredging work on the waterway? Why did they remember the Seaway more than other projects? What insight into the everyday work life and construction problems could they provide that the statistics and newspaper articles overlooked? Finally, what was life like for their families, especially their wives?

While the St. Lawrence Seaway and Power Project has been documented by political historians and engineers (for example, works by William Willoughby and Jacques Lesstrang), their manuscripts address the technological and political aspects of the undertaking and ignore the workers' perspectives. Cold War historians offered a great deal of the ideology and politics of the era (Walter LaFeber and Douglas T. Miller and Marion Nowak), but neglected the social, economic, and cultural changes that occurred in marginal American and Canadian towns.

My study, besides providing a political history of the St. Lawrence project, reveals the human side of the project in the words of engineers, laborers, and carpenters; offers another perspective of the differing work cultures and construction procedures of the Americans and Canadians in the 1950s; and contributes to the era's social and cultural history. The stories of the men and their wives who lived in Massena, New York, and Cornwall, Ontario, for five years supply a wealth of information on changing gender roles and Canadian and American family values. These first-hand accounts provide a vivid commentary of the lives of the children of the Great Depression, who found themselves among the working elite, facing death every day. The St. Lawrence Seaway workers represent a category of construction worker with a unique lifestyle and work ethic that has long been forgotten. These men risked their lives to complete a project of international importance.

Chapter 1 traces the lengthy political history of the project and analyzes how the arguments of U.S. and Canadian politicians regarding the various benefits of the Seaway changed over time and, finally, fit into the political and economic agendas of the leaders of both countries during the early Cold War. Chapter 2 discusses the St. Lawrence Seaway and Power Project from a design standpoint and highlights post–World War II engineering and mechanical breakthroughs. Chapter 3 provides the personal, educational, and work histories of my interviewees. Chapter 4 describes the daily work lives of all levels of workers from management to laborers. Chapter 5 discusses the construction dilemmas of the contractors, supervisors, and engineers who dealt with incomplete plans and inadequate concrete. Chapter 6 addresses the leisure time of workers and their families. Chapter 7 presents the experience of Seaway and power dam workers' wives. The final chapter examines why the project remains so vivid in the minds of the men who constructed it and why it has never gained national recognition.

The St. Lawrence Seaway and Power Project stands as a testament to the unrelenting efforts of politicians, engineers, and workers on both sides of the border. From the beginning of the construction in 1954 to the passage of the first ship on April 1, 1959, workers labored around the clock, braving the elements to complete the project on schedule. Workers, in past histories of the Seaway, have only been given minimal attention. Historians seem to have been more interested in the state-of-the-art equipment used during the construction rather than in the details and hardships of the men who operated the machines. People have been fascinated by the number of ships that use the Seaway without ever wondering how many lives were lost in the attempts to tame the rapids in order to make the river navigable. My study is a tribute to the men who constructed the largest navigational and power project in North American history and participated in the greatest construction show on earth.

The St. Lawrence Seaway and Power Project

1

The Binational Political Debate

AT THE BEGINNING of every semester I ask my collegiate history students to identify the most controversial decision of American congressmen and Canadian members of Parliament in the twentieth century. Their responses often include the U.S. decision to enter World War I and the official commitment of American armed forces to the Vietnam conflict. When I reveal the answer as the fifty-year debate in the United States and Canada over the merit and economic feasibility of the construction of the St. Lawrence Seaway and Power Project on the U.S.–Canadian border, I am met with blank stares. While the majority of my students were born and raised in New York State and some are even from northern New York, they have never visited the numerous locks and dams in their own backyards. Even though William Willoughby in his 1961 book *The St. Lawrence Waterway: A Study in Politics and Diplomacy* documented the political history of the project, the St. Lawrence Seaway and Power Project remains absent from mainstream high school textbooks. More critical events in the twentieth century and the Cold War overshadow the controversy that surrounded the funding and eventual construction of the St. Lawrence Seaway and Power Project. The expense of the undertaking coupled with the questionable long-term use of the waterway prevented the passage of legislation on both sides of the border until 1951.

During the five decades of international debate, Massena and New York State leaders played a positive role in keeping the Seaway project in the national spotlight. Even at times when there seemed to be no support for the power dam and the navigation project, local and state officials wrote letters and spoke to local and national business and political leaders about the project's significance. This dedicated group of supporters realized that eventually the project would loom large on the national agenda.

Even though the fifty-year political debate was the longest in American history, scholars argue that the project's lack of widespread recognition is based on the St. Lawrence River remaining a more crucial aspect for Canada's economic and social development than for America's. As William T. Easterbrook and Hugh Aitken surmised, "Throughout the history of Canada, the St. Lawrence River has served as the major artery of commerce—the axis from which development began, and around which the national economy was organized."[1] The river played a major role in the settlement of Cornwall, Ontario. The area's fertile land and location near a navigable waterway made the region appealing to United Empire Loyalists and the young farmers who relocated to this remote location in 1784. Initially the land near the shoreline was seen as the most valuable as it allowed farmers easy access to water for irrigation purposes and to the St. Lawrence River to transport supplies and food. However, boat captains considered many sections of the waterway dangerous and impassable because of swift currents, shoals, and rapids. Over the next fifty years, Canadian investors and provincial governments continued to dredge new channels and build locks to avoid the rapids surrounding Galop Island, near Iroquois, Ontario, and Niagara Falls. By 1837 over three thousand miles of canals had been completed. These new waterways were further improved under a Canadian government-funded plan that set the standard for canal depth in 1850 and bankrolled the dredging of existing channels based on the dimensions of the newly completed Welland and Seaway canals. This new policy provided a traversable waterway from the Atlantic Ocean to Lake Ontario with a minimum depth of nine feet.

Under the 1854 Reciprocity Treaty, U.S. ship captains gained the right to use the St. Lawrence canals on the Canadian side of the river to transport their freight and human cargo. The original purpose of this agreement was to bolster the Canadian economy by increasing Canadian exports to the United States by guaranteeing the lowering of protective tariffs. The access to these waterways convinced American politicians that constructing comparable canals on their side of the border would be redundant. Therefore, the American improvements on the river remained limited to removing shoals near the Thousand Islands until the commencement of the Seaway and power dam construction in 1954.

Between 1884 and 1905 Canadian government-sponsored programs facilitated dredging in the existing canals to allow ships with a fourteen-foot draft to travel from the Atlantic to Lake Superior. In the 1920s and early 1930s, Prime Minister Mackenzie King adopted a reserved attitude toward the large Seaway project based on the cost of the waterway and provincial disputes over power rights. He did not see an immediate need for the Seaway as Canadian engineers continually upgraded the nation's canal system and rail service to handle public and private transport needs. The completion of the Cornwall Canal in 1934 allowed for the navigation of ships around the Long Sault Rapids in the hopes of encouraging Midwest manufacturers and farmers to ship their overseas freight to Montreal ports via boat rather than sending them by rail to New York City. King's government had already committed funds to update the Welland Canal, to expand the Hudson railroad, and to repay the nation's war debt. L. A. Taschereau and Mitchell Hepburn, the premiers of Quebec and Ontario, the nation's largest provinces, deemed the St. Lawrence Seaway project unnecessary and possibly dangerous to the livelihood of their existing ports and power plants.

The completion of the Cornwall Canal in 1843 and the Massena Canal in 1898 illustrated the ability to harness the power of the current of the St. Lawrence River for electrical production. At this time, Ontario provincial and New York State officials added the argument for a large hydrogeneration plant to their long-standing demand for a deeper seaway. Consequently, American politicians proposed the approval of a combined seaway and power dam in the early 1900s at the time that the Panama Canal was finished. Canadian government officials and engineers had always hoped to build the seven-lock St. Lawrence Seaway exclusively on their side of the border. The power project, however, always needed to be a joint effort based on the fact that the dam site crossed the U.S.–Canadian border. Before World War II, Canadian government officials opposed the power development plan because government-owned power plants supplied enough electricity for all of the country's industries and homes. However, the blackouts that occurred during World War II exposed Canada's lack of a sufficient power supply to support the needs of its growing industry and population. Therefore, the power development aspect of the project surpassed the navigational needs and made the St. Lawrence Seaway and power dam critical elements of the post–World War II agendas of American and Canadian politicians and manufacturers.

At the end of the nineteenth century two events brought the concept of the St. Lawrence Seaway to the attention of American lawmakers: a resolution to Congress and the discussions of a waterway convention in Toronto, Ontario. John Lind, a representative from Minnesota, is credited with being the first American politician to propose building the St. Lawrence Seaway. In 1892 he presented a resolution to the House asking President Grover Cleveland to initiate talks with the Canadian government about deepening the Welland and other canals along the St. Lawrence River. The members of the House Committee on Interstate Commerce recommended the passage of the proposal, but the session ended before the issue made it to the floor. In spite of the quick demise of Lind's proposal, his suggestion introduced the possibility of a St. Lawrence Seaway to House members for the first time. Two years later American and Canadian engineers, politicians, and businessmen from eight Midwest states and numerous provinces discussed canal improvements throughout both countries at a deep seaway conference in Toronto. While attendees explored the improvements of many inland waterways, the majority present viewed the St. Lawrence route as the most important for their national lawmakers to pursue as a joint U.S.–Canadian project.[2]

The inaugural meeting of the International Deep Waterways Commission in Cleveland, Ohio, in 1895 signaled the commencement of the American government's official interest in exploring the merits of constructing a seaway from the Great Lakes to the Atlantic Ocean. From 1895 to the outbreak of World War I, the U.S. government funded several studies by special commissions and engineering boards, whose members concentrated on the transportation aspects of several deep waterway projects. The authors of these studies determined whether such projects offered manufacturers and farmers in the interior of the nation a less expensive means of transporting their products to overseas markets. The Eighty-Third Congress created the Deep Waterways Commission on February

15, 1895. The commission investigated the possibility of improving various rivers from the Great Lakes to the Atlantic, making them accessible to large cargo vessels. In 1897 they concluded that both the St. Lawrence and the Mohawk canals were possible routes and suggested further surveys. The members also appointed a board of engineers. Over the next two years, President Grover Cleveland and Congress spent $483,000 on investigations by the board of engineers. Submitted in 1900, the board's report was one of several before World War II that recommended the construction of the St. Lawrence route at a depth of twenty-one feet.[3]

In an amendment to a 1902 Rivers and Harbors Act, congressmen invited Great Britain to form a joint International Waterway Commission composed of three members from each country to address the Seaway project. As Canada was part of the British Commonwealth, certain issues of foreign policy were still left to the discretion of the British government. The Canadian Order in Council successfully lobbied for the British acceptance of the invitation. In October, the United States selected Colonel O. H. Ernst, George Clinton, and Professor W. Wisner as its commission members. Two years later, the Canadians appointed James Maybee, W. F. King, and Louis Coste to represent their interests on the board.[4] The first meeting of the group took place on June 14, 1905, in Toronto. The work of this organization led to the signing of the 1909 Boundary Waters Agreement between Great Britain and the United States, which stated that "the navigation of all navigable boundary waters shall forever continue free and open for the purposes of commerce to the inhabitants and the ships, vessels and boats of both countries" and created the International Joint Commission of the United States and Canada to deal with boundary water disputes.[5] The 1909 agreement and the commission's findings convinced American government officials, including Charles Townsend, a Michigan senator, of the untapped power and navigation aspects of the St. Lawrence River and encouraged him to initiate hearings on the project in both houses of Congress in 1911. Two years later, the Senate adopted a resolution encouraging President Woodrow Wilson to begin negotiating an agreement with Canadian officials concerning the cooperative improvement of the navigation in the boundary waters of the two countries.[6] However, the Canadian government's failure to respond to the proposal of a joint feasibility study of the deep waterway, and the outbreak of World War I further postponed any action.

After the conclusion of World War I, Bertrand H. Snell, Massena's Republican member of the House of Representatives, brought the St. Lawrence Seaway Project back into the national spotlight. Snell was born in Colton, New York, not far from Massena and he began his business career as a bookkeeper for the Racquette River Paper Company in Potsdam, New York. In terms of his political career, he served as a member of the Republican State Committee and delegate to the Republican National Convention from 1914 to 1944. On November 2, 1915, he was elected to the House of Representatives, where he remained until 1939. During his tenure as chairman of the House Rules Committee and minority leader from 1931 to 1939, he championed the St. Lawrence Seaway Project. On January 29, 1957, New York's twenty-six members of the House of Representatives suggested that the Grasse River Lock be renamed the Snell Lock in his honor. In support of this proposal an *Ogdensburg Journal* editorial proclaimed, "If the name of anyone should be put on the various projects of the St. Lawrence Power and Seaway development, former Congressmen Bertrand H. Snell is far more entitled to the honor than some of the many who climbed on the bandwagon when it became popular. For years Mr. Snell stood alone in Congress as the only member from the Great Eastern United States who supported the Seaway legislation or who spoke of it."[7]

On April 24, 1917, Snell introduced the first Seaway bill. The father of the Seaway, as he became known, stated in a 1955 interview, "It [the Seaway] did not have a single supporter in Congress, other than myself . . . not even any of the senators and congressmen from New York State. For years, I was the butt of all attacks."[8] H.R. 3778 was a bill authorizing a preliminary determination of the feasibility and cost of the St. Lawrence Seaway and Power Project. The document read,

> Be it enacted by the Senate and House of Representatives of the United States of America in Congress assembled that the Secretary of War is authorized and directed to make such preliminary examinations as can be made from available data, without making field surveys, touching the creation of conditions in or paralleling the Saint

Lawrence River from Lake Ontario to the Canadian border, suitable in all respects for navigation by oceangoing ships, including such approximate estimate of cost of improvement as can be predicated on such available data, and an approximation of the amount of power, if any, that would be made incident thereto.[9]

Following the introduction of the Seaway bill to Congress, the International Joint Commission began compiling the most comprehensive study of the navigational aspects of the Seaway. In 1920 the U.S. and Canadian governments presented the group with a list of nine specific questions they wanted answered, including the best route for the waterway; the final construction cost; who would maintain, operate, and regulate water levels after completion; and the predicted overall influence of the Seaway on the commerce and industry of each country. The commission, led by Lieutenant Colonel Charles Keller of the U.S. Army Corps of Engineers and W. J. Steward of the Canadian Department of External Affairs, held a series of forty-six hearings in sixteen states and five Canadian provinces from March 1920 to February 1921. More than three hundred businessmen, private individuals, and farmers testified at the proceedings.[10] Keller, Steward, and the other commissioners also appointed W. A. Bowden, chief engineer of the Department of Railways and Canals of Canada, and Lieutenant Colonel W. P. Wooten of the U.S. Army Corps of Engineers to determine the best way to construct the Seaway and its probable economic impact on both countries.[11] The Commission submitted their 184-page Wooten-Bowden Report to U.S. and Canadian officials in December 1921 and suggested they negotiate a treaty to construct the waterway and mutually share the expense of the project. The International Joint Commission also recommended that a larger engineering board be established to further study the construction aspects of the St. Lawrence Seaway.

Following the favorable recommendation of the International Joint Commission, President Warren G. Harding instructed State Department officials to send an invitation to Canadian political leaders indicating that the United States was ready to begin negotiating an agreement regarding the joint construction plans and budget for the St. Lawrence Seaway. Harding's statement ascertained, "The heart of the continent, with its vast resources in both agriculture and industry would be brought in communication with the execution of the St. Lawrence waterway project. The feasibility of the project is unquestioned, and its cost compared with some other great engineering works would be small."[12] In May 1922, the State Department relayed Harding's message to Prime Minister MacKenzie King. King declined to participate in any discussions at that time, and indicated in his June reply that he wanted to delay any official governmental talks until an enlarged board of engineers conducted a further investigation. He justified his decision to the members of the House of Commons by explaining, "The government has replied to the effect that the present is not an opportune time to consider the report that has been presented and the matter should be allowed to remain in abeyance."[13] King was convinced that the power project was a worthy undertaking, but saw the seaway as having little benefit for Canada and possibly lessening the profits of the Canadian national railways. He also could do little without the support of provincial leaders. The new Ontario premier, Mitchell Hepburn, remained strongly opposed to the project because of its price tag and the lack of evidence that the seaway and power dam would positively impact the regional economy.[14]

President Coolidge and Prime Minister King created the Joint Board of Engineers in 1924 to work with the International Joint Commission in analyzing the many technical issues raised by the Wooten-Bowden Report. Each country also created separate organizations to assess its respective political and economic issues. Coolidge established the St. Lawrence Commission with Secretary of Commerce Herbert Hoover as chair, and King organized the Canadian National Advisory Committee overseen by G. P. Graham, the minister of railways and canals. Both groups were asked to advise government leaders on each side of the border about whether they should participate in the waterway and dam construction at that time. Two years later, the report of the Joint Engineers recommended the project. They estimated the cost at between $625 and $650 million. The project would consist of twenty-five-foot canals, two power dams, five locks of eighty feet in width, and channels in the Thousand Islands, International Rapids, St. Francis, Soulanges, and Lachine sections.[15] The major point of contention between U.S. and Canadian engineers concerned the number and location of the power dams. The U.S. representatives proposed a single dam and power house at Barnhart Island on the American side, while the Canadians wanted two facilities on their side of the border, in Morrisburg and Cornwall, Ontario.

The St. Lawrence Commission presented a favorable report to President Calvin Coolidge in December 1926, and suggested he immediately negotiate a Seaway treaty with Prime Minister King. The document's authors cited the reduction of transport and power costs and the expansion of the number of ships that could use the waterway as the main reasons for their support. President Coolidge submitted the Commission's report to Congress in January 1927, stating, "the time has come to resume the opening of our intracoastal waterways, including the development of the great power and navigation project of the St. Lawrence River. The project should have the immediate consideration of Congress and be adopted as fast as plans can be matured and the necessary funds available. Upon these projects depends much future industrial and agricultural progress."[16] In April, Secretary of State Frank Kellogg sent a note to the Canadian foreign minister in Washington indicating that the United States was ready to sign an agreement with Canada to complete the project.[17] Three months later, Prime Minister King delayed his response to the U.S. request until he received the National Advisory Committee's report.[18] Because of the continued lack of public and political support for the Seaway and Power Project, he wanted concrete evidence that the project would be beneficial to all Canadians and not negatively impact existing ports and railways. King also remained unsure that the navigation aspect of the project was necessary.

Two years later, the National Advisory Committee presented a mostly positive report to King regarding the Seaway design and cost. Their only query was the depth of the canals. The National Advisory Committee argued that the initial dredging of the canals to twenty-seven feet would make the river more accessible to a larger number of ships and eliminate the need for future expansion projects. King again postponed meeting with U.S. officials because he wanted to stall power development in the International Rapids section until there was a national demand for more electricity. He indicated that the project's price tag was too high and the possible loss of freight business for the Canadian National Railways was too risky. King, along with other Canadian officials, believed that it was a strong possibility that the U.S. government would finance and complete the whole Seaway project by themselves, and later allow Canadian ships to traverse the waterway and industries and individuals to purchase power produced by the hydrodam.[19]

The issue of New York State's and the province of Ontario's developing the power element of the project gained support on both sides of the border in the 1930s. After Franklin D. Roosevelt was elected governor of New York in 1930, he campaigned for state power development on the St. Lawrence River. He intensified his effort when General Electric, Dupont, and Alcoa executives joined forces and proposed to construct a power dam and a system of transmission lines across the state to transmit power from Niagara Falls and the St. Lawrence River to New York City. He argued that a private monopoly of power was something that needed to be avoided. In 1907 state officials had refused Alcoa's request to harness the power of the International Rapids section and FDR continued to rally for state control of the region's remaining power sources. New York officials feared that private corporations would develop cheap power to supply their plants and elevate the price for consumers. Hydropower was a natural resource that should be used for the betterment of citizens' lives, not to increase corporate profits. FDR asserted, "I want hydroelectric power development on the St. Lawrence, but I want consumers to get the benefit of it when it is developed."[20]

In March 1930, the New York State legislature established the St. Lawrence Power Commission and an advisory board of engineers to create plans for St. Lawrence power development. On January 14, 1931, they presented their plans for the International Rapids section to the governor. Subsequently, on April 27, the state legislature created the first power authority in the United States—the Power Authority of the State of New York (PASNY)—to oversee the state's power rights and participate in future negotiations between the two nations regarding the joint development of the power dams near Massena, New York, and Cornwall, Ontario.[21] Members of the upper house almost killed this monumental piece of legislation by adding an amendment stripping the governor of the power to select administrators to manage the new authority. Not wanting to relinquish control over government appointments, FDR threatened to veto the bill. In response, North Country residents and organizations telephoned and sent letters to legislators voicing their support of the bill. Additionally, Representative Snell traveled to Albany and convinced three Republicans to vote against the amendment, thus ensuring the passage of the Power Authority bill.[22]

Ontario Hydro officials' efforts to fund and improve the International Rapids section began in 1923. The agency was created in 1906 by the Ontario legislature under the Power Commission Act to construct transmission lines to supply municipal utilities with power generated at Niagara Falls by private companies. The May 7, 1906, bill charged Ontario Hydro with the dual role of corporate regulator and electric power distributor.[23] In its first five decades of existence, the agency erected numerous hydrogeneration facilities across Canada. Ontario Hydro's initial power generation proposal presented to Prime Minister King in 1923 stated that the agency, in cooperation with the provincial government, would erect a dam, power house, and control works in Morrisburg, Ontario, to produce 700,000 horsepower of electricity. However, King contended that he did not have the authority to grant the power rights on the river to Ontario Hydro without further discussions between the provincial and national governments. He also indicated that no decisions would be made about the navigation or power section of the Seaway until U.S. and Canadian engineers completed their investigations.[24]

National leaders on both sides of the border finally acknowledged the importance of constructing the Seaway and power project in 1932 when President Herbert Hoover and Prime Minister R. B. Bennett signed the first Seaway treaty. However, even though Canadians still viewed the idea of a joint Seaway project with apprehension, the political climate became more favorable. The premiers of Quebec and Ontario accepted the navigation and power plans because of the rapid population and industrial growth in their provinces, which were straining the nation's electrical supply. But, with the country in the midst of a depression, Prime Minister Bennett found it hard to convince officials to put the nation further in debt when the current canals easily accommodated the declining national freight traffic. The final report of the Joint Board of Engineers submitted to President Hoover and Prime Minister Bennett on April 9, 1932, convinced the two men of the merits of the project.[25] Hoover voiced his support for the project: "The signing of the Great Lakes–St. Lawrence Waterway Treaty marks another step forward in this the greatest internal improvements yet undertaken on the North American Continent. Its completion will have a profoundly favorable effect upon the development of agriculture and industry throughout the Midwest. The large byproduct of power will benefit the Northeast."[26]

The Seaway and power dam plan that Hoover and Bennett agreed upon bore many similarities to the project that would be completed in 1958. The navigational route contained twenty-seven-foot channels and a two-stage power dam financed equally by the United States and Canada, with the latter getting credit for past work completed on the Welland Canal. The responsibility for the construction of each country's designated locks and dams lay with the contractors employed by each government's appointed lead agency. The United States agreed to dredge channels from Lake Superior to Lake Erie as did the Canadians on the identical stretch of the waterways that lay on their side of the border. The Canadians also consented to improve the Soulanges and Lachine canals. American and Canadian engineers and contractors would complete the power dams, power house, and accompanying channel work near Barnhart and Crysler Islands in the International Rapids section of the St. Lawrence River. Upon the completion of the power project, the U.S. and Canadian utilities and industries would divide the output equally. The Seaway Authority on each side of the border would be charged with maintaining the locks and channels in their section of the waterway. The total cost for the entire waterway and power dam equalled $543,429,000.[27] U.S. senators, however, delayed the vote on the treaty until after the 1932 elections. As schedulers did not place the bill on the agenda, it became part of the Senate's unfinished business when its session ended in March 1933.

President Roosevelt submitted the 1932 treaty to the Senate for a second time in January 1934 after he received the results of an interdepartmental board report outlining the predicted shipping usage of the Seaway and the projected transportation savings. Senators debated the bill for several months on the Senate floor, and to many outside observers, it appeared the bill would pass. But the lack of enthusiasm by Democratic senators and no forceful endorsement from the president led the bill to fall short of the two-thirds majority needed for treaty approval. A few days before the vote, Roosevelt realized the bill would not pass based on the strength of the opposing lobbyists. He still believed that the project would eventually be completed. Roosevelt indicated, "Whether the thing goes through this afternoon or not makes no difference at all. The Seaway is going to be built as sure as God made little apples."[28] The House had voted against the measure 224 to 171 in April of the previous year. Congressmen still emphasized that the damage to the railways, ports, and coal industry outweighed the benefits of a more navigable shipping route on the St. Lawrence and the increase in the electricity offered by the power dam.[29]

In 1934 Massena residents and Republican House leader Bertrand H. Snell began a full-scale campaign to derail the negative statements of Seaway opponents and worked to rally the Senate to ratify the Seaway treaty. Snell stated, "I doubt if there has ever been a project before this country where there have been so many misleading statements that should only be characterized as stump speeches for the purpose of creating an unfavorable feeling against the whole project . . . The average citizen does not seem to understand what we propose to do."[30] According to E. B. Crosby, the head of the Massena Chamber of Commerce, "North Country organizations . . . responded wholeheartedly. The result was that a flood of telegrams began to pour in on the senators. The entire North Country stood together as one man in this movement."[31] Philip H. Falter, president of the Massena Chamber of Commerce, and Walter F. Willson, along with Ogdensburg residents Julius Frank, William H. Cuthbert, and John Van Kennan, traveled to Washington to lobby senators and addressed the Northern Federation of Chambers of Commerce. They argued that opponents of the Seaway had garnered substantial press coverage, while the editors of the Associated Press had not published the opinions of supporters.[32]

Snell shared the benefits of the Seaway and an outline of the project components with his colleagues during a January 17, 1934, session of the

Seventy-Third Congress. He indicated that FDR supported the construction of the Seaway because of its national importance and far-reaching benefits. Snell maintained that once opponents understood the actual cost of the Seaway and its long-range prospects, they would clamor for its immediate completion. In his address he also described the construction of the project and argued that there was nothing partisan or regional about the proposal. On the contrary, he characterized the interests of opposition leaders as sectional and local. Snell asserted that, in the past, the policy of federal officials had been to develop all viable internal waterways, and indicated that every aspect of the project from the engineering to the commercial value had been studied by experts. In response to opponents' criticism of the project's costs, he emphasized the affordability of the Seaway, based on the Canadian government's willingness to split the cost with the United States. Snell continued to counter Seaway opponents in the concluding lines of his speech: "I am honestly of the opinion that it is a good, clean, straight agreement between two nations, and it will further cement the friendly relations between our neighbors on the north and ourselves, and from any study that I have been able to give it—and I admit I have spent some time looking it up—I fail to see one honest, intelligent reason why we should not join with Canada in concluding the treaty and making this development, which is the largest single development on the American continent."[33]

The lingering Great Depression stalled the St. Lawrence Seaway legislation. Although the Seaway presented a feasible project for the employees of the Works Project Administration, its funding still needed to be approved by Congress. During the Depression, Massena mayor Thomas Bushnell worked to keep the Seaway project on the minds of local and national lawmakers. He possessed expert knowledge of the economic benefits of the project and could also address the actual construction elements of the Seaway and power dam, given that he had begun his career as a licensed engineer and land surveyor on the Massena Canal project. As mayor of Massena from 1931 to 1935, Bushnell had traveled with other area politicians and business leaders to the Hotel Franklin in Malone, New York, to encourage Warren Thayer, a state senator from Chateaugay, to cast his vote in favor of the Power Authority bill. In 1931 local Democrats and Republicans considered him a strong contender for one of the trustee positions with PASNY. Even after Governor William Harriman failed to appoint him to the board, he remained a strong proponent of the Seaway and, for a time, became the project's unofficial spokesman. Bushnell presented a speech at a meeting of the Great Lakes Association in Toronto in 1934, expressing his support for the Seaway project and its potential benefits for Massena.[34]

Between 1933 and 1940 several conditions changed in the United States and Canada that brought the Seaway debate back to the national political agenda in 1941. In Canada, one of main opponents of the project during the thirties, Maurice Duplessis, the premier of Montreal, was defeated in 1939 by Adelard Godbout, a Seaway supporter. Mitchell Hepburn, the premier of Ontario, had a change of heart and decided to support the project. King sent word to U.S. officials that he wanted to reopen talks, but once a reporter from the *Montreal Gazette* published King's intentions, opponents stalled his efforts.[35] During this period, American engineers conducted additional design studies of the waterway and national supporters changed their main argument regarding the benefits of the Seaway. The project now became a key aspect of the nation's defense plan, not just a convenient route for ship captains or a cheap source of power for state residents and businesses. This change allowed President Roosevelt to fund Corps surveys of the International Rapids section of the St. Lawrence River in 1933 and 1940 with $1 million from the defense budget.[36] He justified this expenditure by categorizing the waterway as an "integral part of the joint defense of the North American continent." The nation's lack of power limited its production of planes, guns, and other war items at a time when almost all of the enemies of democracy were developing their hydropower resources. According to Roosevelt, the Seaway had ceased to be an opportunity and had become a necessity for defense, transport, and power for increased war production.[37]

In 1941, following power outages during World War II, New York State officials cast aside their former opposition and instead voiced their support for the Seaway and Power Project at hearings held by the River and Harbors Committee in the U.S. House of Representatives. Governor Herbert H. Lehman addressed the economic and national security aspects of the project. He indicated that the nation was facing a shortage of aluminum, equipment, steel, and shipping and that the Seaway would alleviate these problems. Lehman believed that the national benefits overshadowed the argument of opponents who saw the Seaway as having only sectional or local benefits. The governor argued, "We must utilize every means to contribute to the economic strength and security of our country and to protect our way of life."[38] He continued his analysis of the project, emphasizing that its detrimental effect on ports and transport facilities was unsubstantiated and that lawmakers could not afford to sacrifice America's security based on the selfishness of business owners.[39] Lehman agreed with supporters that the Seaway was not only useful and vital but would contribute immensely to the safety and well-being of the entire nation. New York mayor Fiorello LaGuardia added that he would not support any project that he thought would negatively affect the port of New York. He stated that the power project was "as important as water and sunshine and no group of people have the right to monopolize sunshine or water."[40]

At the federal level, President Roosevelt attempted to gain passage of the Seaway plan in 1941 in the form of an executive agreement to avoid the two-thirds majority vote required for treaty approval. While Prime Minister King had signed the document, he decided not to pursue parliamentary support until Roosevelt gained congressional approval. The U.S. lawmakers delayed the vote on the measure because of the bombing of Pearl Harbor. Many feared a project of this magnitude would subsume workers and materials needed for the war effort.[41] Roosevelt tried financing the project with discretionary funds, but came up short. He asked Robert Patterson, the undersecretary of war, for $50 million from the War Department budget. Patterson did not view the Seaway and power dam as an urgent wartime project and declined to divert the funds.[42] Regardless of the war, both Ontario Hydro and the Corps sent employees to Massena and Cornwall in 1942 to do test borings for the power dam.

The presence of members of the Corps in Massena convinced a local dentist and a major supporter of the Seaway and Power Project, Dr. Rollin Newton, of the impending federal approval of the Seaway plans. For several decades he had explained the benefits of the Seaway to anyone who would listen. Newton subjected his dental patients, during fillings and root canals, to a fifteen-minute optimistic lecture on the Seaway. *Massena Observer* editor Leonard Prince described in the newspaper his first teeth-cleaning experience in 1928 during which Newton informed him of the reasons why the Seaway needed to be built and assured him that the U.S. Congress would pass the project bill during its next session. Prince admits that he lost hope as the years went by and no construction began, while most Massena natives, led by Newton, remained confident the Seaway would eventually be completed.[43]

As president of the Northern Federation Chamber of Commerce, Newton continued writing to senators and other officials encouraging them to pass the Seaway bill. In 1945 he turned his attention full time to convincing state and federal lawmakers of the urgency of passing the St. Lawrence Seaway legislation. He composed a letter to Senator George Aiken of Vermont, a longtime supporter of the Seaway, and attached a press statement sent to the *Rochester Times Union, Syracuse Post Standard,* and *Buffalo Evening News.* The opinion piece outlined the importance of developing the hydroelectric potential of the St. Lawrence River and challenged Seaway opponents to a public debate. Newton asked for support from unions based on the estimated employment of 100,000 people during construction at the site and in related industry. He also criticized project opponents for producing a weak, negative propaganda not supported by facts, but fueled by fear and selfish interests. Newton argued that it was an opportune time to use the press to educate the public about the benefits of the project. He indicated that a U.S. Department of Commerce survey of the Seaway project uncovered no ill effects to railroads, and cited cheap hydropower and transportation for industries and increased employment in manufacturing. Newton concluded that once people knew the true benefits of the Seaway and the power project they would push for their immediate construction. He also wrote to President Roosevelt and asked him to consider approving the Seaway under the National Defense Act. Newton strove to inform national and state politicians, as well as New York citizens, of the benefits of the project to the state and the nation as a whole.[44]

In 1945, former assemblyman and Massena resident Grant F. Daniels spoke at the Watertown Kiwanis Club regarding the benefits of the Seaway for industry, agriculture, and labor. As cosponsor of the Graves-Daniels resolution in the assembly, he had pushed New York lawmakers to go on record with their unanimous support of the project. Daniels' presentation followed the speech of Dr. Lewis Sillcox, who had addressed the club the previous year about the negative impact of the project and condemned its regional benefits. Daniels wanted to present the advantages of the waterway to the general public and dispel the myths of critics. According to Daniels, "The day will surely come when public interest prevails over private interest."[45] Daniels also pointed out that the opposition's argument that the Seaway could be constructed as two separate projects—power and navigation—was unrealistic, as one could not function without the other. The Seaway offered New York citizens and businesses access to cheap power and a new navigation route to the Atlantic. Both served America's Cold War initiative by providing greater international trade and services, increased domestic employment, expanded rural electrification, and a foundation for lasting world peace. Daniels concluded that he believed that the Seaway would eventually be built because of its overall benefits and the long-standing American tradition of not allowing special interests to dictate public policy. The threadbare and selfish arguments of the opposition would eventually be exposed, he prophesized, and lead to their discredit.[46]

In the midst of lingering American political delays, Canadian Prime Minister Louis St. Laurent and Minister of Transport Lionel Chevrier aggressively rallied Parliament to approve the project and suggested Canada complete the waterway independently in the late 1940s and early 1950s. The Seaway project became increasingly important to Canada in the post–World War II era. War production had catapulted the nation into its third industrial revolution. A concentration of businesses and residents along the St. Lawrence River, accompanied by an enlarged national population, also caused power shortages in Ontario between 1945 and 1955. To combat this problem, Ontario Hydro officials opened fourteen new hydrodams and increased their annual output of electricity from 1,852,000 to 4,229,100 kilowatts.[47] The International Rapids section of the St. Lawrence River remained the country's last untapped power source by the 1950s. American entrepreneurs also had rediscovered iron ore fields in Labrador, Quebec, whose content they planned to mine and sell to the owners of the Great Lakes' steel mills.[48] The owners of the ore reserves had explored various land transportation options for their product, but the heavy material was most efficiently transported by boat. However, the present fourteen-foot-deep canal system on the St. Lawrence River was too shallow to accommodate the large freighters used to haul iron ore.[49]

In 1949 Prime Minister St. Laurent met with President Harry Truman and emphasized how the shortage of electricity in Ontario made the immediate completion of the power dam segment a priority in his country. Chevrier stated in his memoir that Canada could not afford to build the Seaway and the power project alone until after 1950 because of the nation's war debt and the Great Depression. However, after World War II St. Laurent and Chevrier become confident of the possibility of the completion of an all-Canadian Seaway.[50] During a discussion between the two in Ottawa in 1950, St. Laurent affirmed, "We should build the Seaway alone. I think the Americans should be made aware of our determination to get the Seaway built."[51]

On December 4, 1951, Lionel Chevrier introduced to the House of Commons the St. Lawrence Seaway Authority Act and an act authorizing Ontario Hydro to construct a power dam in the International Rapids section. In his speech he outlined the power production project and emphasized the economic impact on the transportation costs for iron ore miners and prairie grain farmers. The bills passed the House unanimously. Even the three opposition leaders, George A. Drew, M. J. Coldwell, and J. H. Blackmore, in a rare move voted favorably for the bills. The main reason for their support was not the financial benefits that their constituents would reap from the completed project, but their desire to keep the Seaway all-Canadian and to shut out the Americans.[52]

The U.S. Congress rejected the Seaway bill three times, in 1944, 1948, and 1952. Regardless of the lack of federal support in 1951, Dr. Newton continued his unrelenting campaign for the Seaway. He wrote to twenty-seven members of the House Public Works Committee and President Truman to ask for support of the Seaway project. Newton appealed to politicians to "give favorable approval to the St. Lawrence Seaway and Power Project. For 30 years we in Northern New York have been hoping and expecting that some favorable action could be taken by Congress in getting this development started."[53] Newton emphasized that Canadian officials were tired of waiting for American approval and were ready to widen and deepen existing canals on their side of the border. He received several responses, including one from Senator Irving Ives of New York, a longtime opponent of the project. Ives indicated that he was impressed by opponents' arguments but remained open-minded. Senator Alexander Wiley of Wisconsin, a proponent of the Seaway, wrote that he saw it as an absolute sin that the project had been delayed, but unfortunately still saw strong opposition from southern and eastern ports and railroads.[54] In October Newton wrote to President Truman to express happiness over his support for the Seaway. He received a response indicating that the day his letter was received the president was speaking with the Canadian prime minister concerning the Seaway and power project. Both agreed on the importance of the project for the American and Canadian economy and security. Truman hoped for congressional approval of the navigation and power project, but indicated that he would support an all-Canadian waterway.[55]

Senator Alexander Wiley of Wisconsin and Representative George Dondero of Michigan led the final congressional campaign to gain the passage of the St. Lawrence Seaway legislation. According to William Willoughby, the Seaway bill passed in the 1950s because of the changing role of the United States in terms of national and global defense, international trade, and foreign relations. In terms of defense, many national officials saw a correlation between the military capabilities of a country and its internal waterways and waterpower.[56] Secretary of State Dean Acheson indicated in a draft of an article for a 1945 edition of the *Democratic Digest* that there was a direct relationship between a country's ability to produce and use energy for production purposes and its ability to prosper in peace and defend itself in war.[57]

In a February 20, 1951, statement to the House Public Works Committee, Secretary of State Dean Acheson indicated, "Events in Korea demonstrate we must develop further our military and economic strength. In order to bear these burdens with assurance of success . . . we must select measures we can take on now which will assure the greatest returns in increased strength and security later on. I believe the Seaway and Power Project is outstanding in this respect."[58] The Seaway also provided an internal route for merchant and military ships to travel unthreatened by submarine attack in times of war and offered numerous ports for ship-building and repair. This route was important to military officials because of the number of men and materials lost during World War II when North Atlantic convoys were attacked by enemy submarines. The war also convinced U.S. political and military officials of the need for a close trade and military relationship with Canada. In a new era of good-neighborliness between the United States and Canada, military leaders believed the two countries formed a single defensive unit. In a 1951 memo from Senator Alexander Wiley of Wisconsin to his fellow senators, Wiley stated, "The St. Lawrence Seaway will add substantially to the economic strength of Canada and the U.S. When the chips are down, the strength of Canada and the U.S. may be all we can depend upon to resist the spread of Communist aggression."[59]

Representative George Dondero introduced House Joint Resolution 104 authorizing the construction of the Seaway on January 9, 1953. Two weeks later, Wiley introduced a similar bill, S589, in the Senate. In April and May 1953, Wiley, the chairman of the Foreign Relations Committee, held hearings for five days with testimony from Deputy Chief of Army Engineers Brigadier General B. L. Robinson and N. R. Danielian, the head of the Great Lakes–St. Lawrence Association and former secretary of commerce. The hearings concluded after a final day of opposition testimony led by leaders of the railroad industry. Wiley, under the advisement of President Eisenhower, delayed the introduction of his bill to the Senate until a report from a six-man presidential committee made a final assessment of all aspects of the project and the National Security Council gave its final approval. Following the receipt of both of these documents, Eisenhower sent the bill to the Senate on January 7, 1954, recommending its passage.

The U.S. Senate and House of Representatives approved the Wiley-Dondero Act on January 20, 1954, and May 13, 1954, and President Eisenhower signed the bill on May 13, 1954, ending the nation's oldest political debate. The main points of the agreement authorized the completion of the U.S. section of the Seaway in the interest of national security. The legislation also differed from the 1932 treaty and the 1941 executive agreement in terms of cost. The navigation and power project had a smaller price tag, as New York State and Province of Ontario officials had undertaken the funding for the power segment on the International Rapids section. The final proposal also lacked plans to deepen the waterway to Duluth, Michigan, in the foreseeable future. These measures, and the charging of tolls, made the project self-liquidating, slashed $100 million from the project budget, and won over those congressmen who had long opposed the project. The power project also offered a new source of power for owners of plants involved in the expansion of the U.S. defense industry. Finally, the new waterway provided an alternate transportation route for supplies to American allies and to American inland ports in the event of an enemy attack. At the Public Works Committee hearings in February 1951, Defense Mobilizer Charles E. Wilson, a former opponent of the Seaway project, stated that the new transportation route and the additional power supply would contribute to the development of Canada's huge resources and to making her an even stronger partner of the United States, whether in peace or war.[60] The passage of the 1951 and 1954 Seaway bills led to the construction of the largest hydroelectric dam in North America and a waterway comparable to the Panama Canal.

2

The Project

IN 1954 CONSTRUCTION COMMENCED on the five sections of the St. Lawrence Seaway and Power Project, three of which were located exclusively in Canada. The completion of the project required the cooperation of New York State and Ontario provincial officials along with four lead agencies, the federal governments of Canada and the United States, and their contractors. Each section included locks, channels, and dams whose construction and eventual operation had to be coordinated among many entities. The overall goal of the project designers and contractors historically remained unchanged: to create a navigable waterway from Montreal, Quebec, to Lake Ontario, where lakers would be raised 225 feet from Montreal to Lake Ontario; to tame the existing rapids for power production; and to bypass the hydro and control dams with locks and canals. According to John Brior, author of *Taming of the Sault*, "The St. Lawrence Seaway project was the 8th wonder of the world, a power-waterway development so great that it defied comprehension, except by the few who planned and built it."[1] Carleton Mabee added in *The Seaway Story*, "The job was diverse. It meant seizing land; lifting bridges; moving houses, railways, and factories out of the way; it meant building canals, dikes, dams, and locks; it meant replanning old towns and creating entirely new towns. The job was daring too. The St. Lawrence water was swift. Valley soils and people were stubborn."[2]

The Canadian aspects of the project included the Lachine and the Soulanges sections and Lake St. Francis. The thirty-one-mile Lachine section stretched from Montreal to Lake St. Francis and included ten new bridges and alteration of others, channel dredging, the construction of two locks—St. Lambert and Cote St. Catherine—and the La Prairie Basin Canal. The channel offered ships an alternate route around the Lachine

Rapids and the locks raised vessels fifty feet to the level of Lake St. Louis. The new route, instead of following the north shore as the original Lachine channel had, used the south shore, which was a much longer route. The main rationale was to avoid passing through busy Montreal Harbor with the swift St. Mary's current.[3]

After Lake St. Louis, contractors created the sixteen-mile Long Soulanges section that housed the Beauharnois Canal, the Upper and Lower Beauharnois Locks, and the Beauharnois Power Plant. The two locks lifted ships another eighty-five feet and allowed pilots to avoid the Cascades, Split Rock, Cedars, and Coteau rapids between Lake St. Louis and Lake St. Francis. The power plant took advantage of an eighty-foot drop. Workers completed two supporting dams that regulate the flow along this section. The last Canadian segment, the Lake St. Francis section, consisted of twenty-nine miles of dredging and terminated at Cornwall.[4]

The forty-four-mile International Rapids section reigned as the most comprehensive segment of the project and presented the greatest challenges for American and Canadian Seaway contractors. The International Rapids section of the project began at the head of Lake St. Francis and extended to Ogdensburg, New York. This part of the river consisted of both American and Canadian locks and dredging projects and required a high level of coordination. The main facility constructed in this area was the mutually built and maintained Robert Moses–Robert H. Saunders Power Dam. Other supporting elements included the Massena Intake and the Long Sault Dam. Contractors dredged the Wiley-Dondero ship channel to allow ships to completely avoid the power operation. The Corps designed and supervised the construction of the Snell and Eisenhower Locks, which lift ships forty feet from Lake St. Francis to the power pool. Planners designed both locks with 1,650-foot guide walls and 115-foot lock walls, for a price of $40 million. The maximum ship length for both locks was 768 feet. Workers also created fourteen miles of dikes to contain the head pond and a water storage area for operation of the power house. The American contribution to the Seaway construction concluded with the removal of shoals in the sixty-eight-mile Thousand Islands section.[5]

Project planners realized that the dams and locks would not be completed on time and on budget without the adherence of contractors and their employees to a stringent time schedule and the avoidance of excessive cost overages. When the PASNY board of trustees and the Corps leadership selected the men to oversee their sections of the St. Lawrence Seaway and Power Project, they chose individuals with experience and unfettered determination. PASNY appointed Robert Moses as its administrator. Even though he spent most of his time in New York City, his periodic visits to Massena and his land acquisition policies made him feared and respected.

Robert Moses, a New York City native, graduated from Yale University in 1909 and was eminently qualified to head PASNY during the construction of the Robert Moses Power Dam and its supporting facilities. His experience in terms of public works construction and his hard-nosed attitude made him the perfect candidate to deal with a stringent construction schedule and reluctant contractors. During his time at Yale he garnered the respect of his swim teammates and fraternity brothers and became known not only as an idealist, but as a fierce competitor. Many of his classmates later recalled that they anticipated his eventual success in business or politics. Over the course of his five-decade public service career, he served as New York City's park commissioner, construction coordinator, and member of the City Planning Commission.[6]

In 1914, New York City mayor John Purroy Mitchel employed Moses to reorganize the municipal civil service system. Mitchel wanted public employment and promotion to be based on merit instead of the deeply embedded American practice of patronage that allowed elected political officials to award federal and state jobs to men who had contributed to their political parties and campaigns. Moses's assignment brought him into conflict with Thomas J. McManus, the leader of Tammany Hall, and with those of other political machines in New York City who did not approve of his efforts that weakened their political control. McManus and his cohorts gained Moses's removal from the Mitchel administration and stymied his ability to attain another state job for four years.[7]

When he returned to public life following Al Smith's election in 1918, he realized that his ideas and dreams of improving society and people's lives meant nothing without the money and the power to make these programs a reality. Beginning in 1945 during his tenures as parks commissioner, construction coordinator, and member of the City Planning Commission, he physically reshaped New York City, building bridges, expressways, parks, and public housing. Moses demanded funding for projects that some found frivolous and exercised complete control over their design and construction. Often referred to as a bulldozer, he squashed any existing structure or individual who stood in his way. Moses often disagreed with engineers who classified certain projects as too difficult to design or construct, including the opponents of the international power dam on the St. Lawrence River.[8]

When Moses arrived in Massena as the chairman of PASNY in 1954, he possessed career experience, the confidence to complete a difficult and controversial project, and a unique insight into the design elements of the power dam. Ironically, in 1918 his brother Paul had been assigned the job by his employer, Consolidated Edison, to find a way to harness the power of the St. Lawrence River. Even though Robert had discussed the power dam design with his brother and knew of his expertise, he never offered him an engineering position on the project even when faced with a shortage of qualified dam design specialists. He had never made an effort in the past to help his brother obtain employment on the numerous public works projects he supervised in New York City, and the Massena Power Project would be no different. Moses's employing his brother would have seemed to support the traditional political practices of patronage and nepotism that he had historically opposed.

Moses ruled the power dam construction with an iron fist and with little outside interference. He had received permission from the other administrators of the Power Authority and Governor W. Averill Harriman to run the American side of the St. Lawrence Power Project construction as he saw fit. Moses used this dictatorial power to force his contractors to complete the dam project and its related facilities in a timely and cost-effective manner. The trustees and the governor only intervened when concern reached the governor's office that Moses was authorizing the demolition of many cottages and family homesteads on the shorelines of the St. Lawrence and causing a public backlash. Even after James E. Truex, one of Harriman's aides, visited the area and confirmed the complaints, a letter from Moses to the governor justifying his action put an end to the scrutiny by state officials.[9]

Jack Bryant

I think much of the credit for things running so smoothly and for the project being built goes to hard-nosed Robert Moses, who headed PASNY. His theory was that he would suffer no fools. He would rook no obstacles and he charged full speed ahead. If someone said, "You can't do that, we have to wait for the permits!" he would just ignore him and go ahead. This attitude resulted in a huge number of changes and extra costs, but it was also the reason why the power dam got completed on time. If there was a problem regarding the cost or pace of the construction, it often got resolved by Robert Moses. It was his favorite ploy to say that he had the authority to make all final decisions because the bondholders who had funded a lot of the project had given him the power to protect their financial investment. It was always in the interest of the bondholders.

Power Authority designers also grandiosed the power plant, causing the price for that structure to escalate. Moses changed the design of the power plant structure into a national monument, adding a penthouse with a museum for visitors and conference rooms for the Power Authority trustees' annual meeting. A guest house was also added for the trustees. It was the first place I had ever seen acoustic materials installed between individual rooms. PASNY officials had meetings with plywood consultants to select the actual flitches of plywood to surround those meeting rooms and the features of the mural. I was enraged when I saw the extra cost of that. The lid was off when they went after stuff like that. On the exterior of the building Robert Moses insisted on a type of sand from Michigan that was white to hold the white sandstone facade. The suppliers brought it in and it was beautiful. However, the bricks would not sit in the sand because of its rounded granules. So the masons had to take it out and put in ordinary mortar and trowel it way back. Then they went back and put in a little front line of Michigan sand mortar. PASNY had a perfectly serviceable power plant design, and they decided to make it into a monument to Robert Moses. Engineers also altered the size of the turbines and the operating equipment to reflect the latest technology. Planners initially designed the power plant structure as a utilitarian structure and instead it became an elegant monument.

1. In September 1958, workers put the finishing touches on the retaining walls and curbs at the main entrance of the visitor's center at the Robert Moses Power Dam. Courtesy of Alfred Mellett and PASNY.

Jim Cotter

The big boss was Robert Moses, who was chairman of the New York Power Authority as well as a myriad of other New York City authorities. I recall a letter that Bob Moses wrote to the chairman of the board of U.S. Steel. The letter started out asking how his wife and family were and after that introduction he lit into him about the delay in the shipping of steel to the power dam contractors. It was kind of [an] interesting letter. It illustrated how Bob worked. He threw his arm around someone and then hit him in the jaw.

Contractors who could not meet their construction deadlines because [of] a lack of workers or supplies could not ask for an extension. Bob Moses had made promises that his contractors would complete the power project by Inundation Day. Moses "the Bulldozer," as we called him, told his employees and contractors that the dam would be completed on time or so many heads would roll that it would plug up the St. Lawrence River.

Moses was a very important, powerful, and assertive person. I have read that Bob Moses was the one person that Franklin Delano Roosevelt wholly despised. And you know FDR was a pretty gentle person.

On one of Moses's visits to Massena, he came to the Uhl, Hall, and Rich office. He made a trip to the toilet. On his return he announced, "Paper towels are the invention of the Devil!" Within the hour, Bill Klepser, the office manager, was racing to Massena to get some high-quality cloth towels before the chairman's next visit to the toilet.

Alfred Mellett

I remember one day the Uhl, Hall, and Rich office staff was prepared for a visit from Robert Moses. They had set up a special dinner for him in the conference room. But when he arrived he said a brief "hello" to everyone and came over and asked me if I wanted to go out and have dinner. He was the one who hired me as the official photographer for the Power Authority on the project and gave me a chance to hone my photography skills, which before had just been a hobby. Moses wanted to get the dam constructed and start producing power. He wasn't there to make friends.

Joe Marmo

I thought Robert Moses was nuts, but it turned out he was right. He did not want to slide the schedule, so he always pushed for various segments of the project to be finished on time. I thought it was kind of stupid because it was costing us money. We never finished the design of any aspect of the power dam before we issued the contracts for bidding. He always had the construction stiffs going twenty-four hours a day, seven days a week. When we finished on time in four years, he was now getting power revenue, but he had paid huge contractor claims for changes and overages. It was kind of a trade-off.

Thomas Airis assumed the duties of the director of the Corps Area Office in Massena and brought with him the knowledge and the reputation to accomplish the difficult task ahead. Airis mollified the criticisms of doubting politicians who waited in the wings for the project to falter, dealt with frustrated contractors struggling with incomplete plans and difficult weather and soil conditions, and motivated and maintained a workforce that was overworked and far from home.

Robert Carpenter, one of Thomas Airis's employees in both the Buffalo District Office and the Massena Area Office, described Airis's stellar career.

Robert Carpenter

It was an honor and privilege to serve under the leaders of the Corps in the 1950s given the assignment to conquer the majestic St. Lawrence waterway. Such a leader was Thomas F. Airis, area engineer for the Buffalo District, carrying out all of the traditions of the Corps of Engineers, U.S. Army. He got things done; he was capable and understood the job; he was well liked in the community. Airis was a soft-spoken individual with a vibrant note of authority.

Airis was born in Eau Claire, Wisconsin, in 1906 and attended Eau Claire High School. He began his studies for a bachelor of science degree in civil engineering at the University of Wisconsin in 1926. After graduating in 1929, Airis started his career with the Corps in the Detroit District, where he was in charge of the construction of the Alpena Harbor, Saginaw River Improvement, Monroe Harbor, and other river and harbor work. He remained in the Detroit District until July 1943 when he was drafted into the military. During the war years, he began as a captain and in 1945 he attained the rank of lieutenant colonel and served as the troop commander and staff officer for the Okinawa invasion. He constructed the port of Naha and remained in Okinawa until 1947, when he was discharged as a full colonel and returned to the Detroit District.

Soon after Airis was transferred to Athens, Greece, where he served as the area engineer for a multimillion-dollar rehabilitation program, which included the reconstruction of the Port Salmonica and the construction of airfields, road reconstruction, and numerous railroad bridges in northern Greece. In 1949 he returned to the United States, but remained only until February 1951 when he was transferred to Saudi Arabia. As the area engineer, he was responsible for the construction of the Dhrahran Airfield and other projects under the Middle East Division located in Tripoli, Libya. After a two-year tour of duty, Airis once again took up his duties as construction management engineer in the Detroit District, where he remained until being named area engineer for the construction of the U.S. portion of the Seaway. As agents for the St. Lawrence Seaway and Development Corporation, the Corps was in charge of constructing two locks and the ten-mile ship channel.

Kenneth Hallock

During the construction the U.S. Army Corps of Engineers reopened the area office in Massena under the direction of Tom Airis. This office housed structural engineering, concrete, mechanical, electrical, and soil experts. On each construction site, the Corps employed a residential engineer, shift engineers, and inspectors who supervised the contractors' work and were in charge of all aspects of daily construction. As an employee of the area office, I went out to the Eisenhower Lock, the Snell Lock, the intake structure, and the Cornwall International High-Level Bridge, and provided whatever services or advice the supervisors needed. The biggest single thing that I did was work with the contractors in preparing lift drawings. Lift drawings showed the location and dimensions of each concrete pour and the required mechanical equipment to install the concrete. The plans showed all of the features that contractors had to incorporate in each pour. There were a couple thousand of these prepared for each lock. I also oversaw the installation and testing of the gates, the gate machinery, and the fenders put across the locks to prevent a boat from slamming into the exit doors. Lastly, my co-workers and I made sure that the electrical systems for each lock were 100 percent functional.

Besides being overseen by two of the most experienced men in public works con-struction, the St. Lawrence Seaway and Power Project gained worldwide rec-ognition for several other notable reasons. First of all, as the waterway crossed an international border, American and Canadian contractors not only split the construction of the power dam and dredging responsibilities, but they also needed to coordinate their efforts on all of the sites on both sides of the river. This split in responsibilities required an unprecedented amount of cooperation and communi-cation between the government officials of both countries as well as between their lead agencies and subcontractors.

Second, in the past during the construction of projects such as this, the staff of the divisional offices of the Bureau of Reclamation, the Tennessee Valley Authority, Ontario Hydro, or the Corps had overseen the entire undertaking from the design phase to the eventual maintenance and operation of the completed facilities. In the case of the St. Lawrence Seaway and Power Project, four separate entities took a collaborative role in terms of design and construction supervision. On the Cana-dian side, Ontario Hydro shouldered the burden of completing the power dam section and the Seaway Authority tackled the dredging and the remaining locks, bridges, railway, and road repair. On the American side, PASNY constructed all of the elements related to the power dam, and the Corps supervised the dredging and lock completions. Each entity put their sections out for bid, hired qualified contractors, and guaranteed that their efforts stayed on schedule and dovetailed with the efforts of the contractors on the other sites.

Joe Marmo

The St. Lawrence Seaway and Power Project was really two projects rolled into one: a waterway and a power production facility, which entailed not only the main dam but all of its supporting control structures. The Power Authority of the State of New York (PASNY) controlled the St. Lawrence River for forty miles from the power dam site to the Iroquois Dam. My bosses cooperated with the Corps of Engineers and Ontario Hydro to con-trol the level of the river. It was our goal not to damage or inundate the locks under construction, so that was what made this project so unique. There were so many entities involved, including two countries. When American contractors built the power dam, they had to meet the Cana-dians at the international border, which was right in the middle of the St.

Lawrence River, and everything had to be in alignment. The other half of the project, including the dredging and preparation of the navigational channel and the supervision of the construction of the Eisenhower and Snell Locks, was the responsibility of the St. Lawrence Seaway Development Corporation and the U.S. Army Corps of Engineers.

Kenneth Hallock

My personal opinion was that it was a very smooth relationship between the four agencies involved—the Canadian entities, the Seaway, PASNY, and the Corps. Obviously the administrators and the engineers from each agency came with their own plans for each structure and how they thought it should be constructed. When these ideas did not agree with the other parties involved, negotiations took place and compromises were made. Sometimes these negotiations were amicable and sometimes they were confrontational. The biggest disagreements were over the time schedule and whether certain dredging projects were necessary. There was also a big concern that each side of the power dam was being built differently. Frequently these big projects like the Panama Canal involve different agencies and entities. I imagine the Corps would have preferred to have had complete control over the design and construction phase of the entire project and eventually operate the completed facilities as they did on their other lake projects. The more control you have, probably the better job you can do. Everyone wants to control their destiny. All in all, with all of the different contractors and agencies involved, it wasn't without hitches, but it went pretty smoothly.

John Bryant

The organizational and the political complexity of funding, designing, and constructing the Seaway and power dam is hard to understand. Two federal authorities, the St. Lawrence Seaway Authority and the St. Lawrence Seaway Development Corporation, oversaw the completion of the waterway including all of the channel dredging and supporting facilities. In terms of the power project, New York State financed and put out for bid half of the main power dam and the Province of Ontario addressed the other side. Ontario Hydro and PASNY oversaw the contractors and construction of the various components of the power project for Ontario and the state of New York, respectively. All of the members of these organizations had to get along with their counterparts across the river and constantly convey to one another their progress and difficulties. The heads of the district offices also had to get along with their vertical counterparts.

The contractors and many of their higher level employees also had large egos. I remember Elwyn Simpson, the project manager at the Long Sault Dam, talking about a bid preparation meeting between the estimators of the various contractors involved in a joint venture. At these gatherings, engineers came up with a price to submit to the Corps of Engineers for the construction of one of the locks. Simpson described how they sat around the table with their hands over their papers and made each other guess the cost for each element of the lock from the common excavation to the concrete. The final bid amount was the result of hours of swapping and horse trading back and forth over how much each contractor was going to receive for his aspect of the construction. The estimators were particularly concerned about areas where they had lost money on a previous job. They wanted to make sure they charged a higher price for their services up front. Somehow the contractors' bids got put together and the Corps awarded these joint ventures huge sums of money to complete the locks, dams, and dredging on each project site.

Keith Henry
Eventually Ontario Hydro and the Power Authority of the State of New York were named the entities to carry out the power development. Canada formed the St. Lawrence Seaway Authority and the U.S. set up the St. Lawrence Seaway Development Corporation to construct their respective navigation works. The development of the International Rapids section between Prescott-Ogdensburg and Lake St. Francis was done under the order of the International Joint Commission (IJC), which was concurred on by the governments of the United States and Canada. The governments also established a Joint Board of Engineers to oversee the design and construction of all of the works. The IJC established the International St. Lawrence River Engineering Board and the International St. Lawrence River

Board of Control. These administrative arrangements were in place by 1954 and continued until the projects were completed and in operation.

Under IJC's jurisdiction, the power entities were required to build the Moses-Saunders Power Generating Station near Cornwall, the Long Sault Dam, and the Iroquois Regulating Dam at Iroquois. In addition they were required to do the necessary excavation of channels for velocity reduction for both ice cover formation and for navigation in the International Rapids section. The alignment for navigation had to satisfy the Seaway entities in the reaches between Galop Island and the dams.

The Seaway entities were required to build the navigation locks and channels downstream of the International section of the river as well as the Iroquois Lock. The Canadian Seaway Authority built the locks from Lake St. Peter below Montreal to Lake St. Francis and also the Iroquois Lock. The American Seaway Development Corporation was responsible for the Eisenhower and Snell Locks and the power house forebay from Lake St. Francis to Lake St. Lawrence.

Before the commencement of construction, the Corps engineers struggled to update their 1941 report. This document included the preliminary designs of each dam and lock and initial soil and subsurface testing and investigations. Junior engineers along with soil experts strove to supply contractors with drawings and test boring results so they could submit informed bids for each site. However, because of time constraints a lot of data on materials that needed to be dredged and on the test borings remained vague and sporadic. The information served as a starting point for contractors at the locks and dams but the final plans remained works in progress. The Buffalo District Office engineers along with their counterparts at Uhl, Hall, and Rich, the supervising contractor for PASNY, completed the detailed drawings a few weeks in advance of the construction schedule and forwarded them to the appropriate contractors. Office engineers in Massena along with contractors assessed the merits and viability of design changes as they arose.

Kenneth Hallock

In 1941 just before the United States entered World War II, the U.S. Army Corps of Engineers had originally designed the Seaway. These plans were incomplete and were part of a very hurried design operation that got under way to construct and build the facility for war use. American politicians argued that the new waterway was needed to transport a new source of Canadian iron ore to the factories making war supplies in the Great Lakes. These plans were very sketchy, and the idea was that these drawings would be a rough guide. The more detailed design of many of the features would actually be done during construction. Then a new Corps district office was created in Massena, New York, to conduct preliminary survey work. But when the Roosevelt administration decided not to proceed with the construction, the office was closed. These prewar blueprints were our starting point at the Buffalo Division Office of the Corps when my colleagues and I undertook the new designs in 1953. Initially the Corps thought that we would run the show, but in the end the Seaway Corporation was our boss. However, even to complete the new plans and perform our supervisory role, my bosses needed to enlarge our office space and staff in the Buffalo District because that was the biggest project that division of the Corps had ever been a part of.

Robert Carpenter

My first responsibility with the Corps began prior to the beginning of construction. Reuben Haynes, the Corps soil expert, was looking for an assistant to run his errands and meet his other clerical needs. I was hired for the position and they put us in a room in a long wooden building at the end of Bridge Street in Buffalo with a lot of big tables to update all of the old 1941 Corps drawings. We also compiled all of the preliminary subsurface investigation results on the various dredging and lock sites before the contract department put the projects out for bid.

Joseph Foley

When the Buffalo Division of the Corps knew they were going to design the Seaway locks, the administrators brought a man in from Pittsburgh, Pennsylvania, who had done a lot of lock design on the Ohio River. He was assisted by the chief of design, a man from Alabama who had also overseen several lock projects. Those two were probably the only ones who had any real experience in lock design. They knew what they were doing and they were good people. Most of the men I worked with had no experience. My fellow engineers and I had the Corps' basic lock design manuals that pretty much told us what we had to do to draw the plans, so it wasn't that difficult. There wasn't anything that was all that complicated. My co-workers and I designed both locks from scratch. Even though we had the old 1941 report, I don't think we ever looked at it. The requirements of and the materials used to constructed these new locks were totally different than what was expected prior to World War II, so it really was like starting over.

After the provincial government of Ontario designated Ontario Hydro as the lead agency on the Canadian side of the power dam, Hydro administrators decided to defer the construction phase of the hydrodam to outside contractors. However, Hydro engineers filled the roles of site inspectors and design consultants. In May 1953 Hydro assigned 177 engineers to conduct depth surveys and begin construction of the models of the entire project before U.S. and Canadian government officials sanctioned the project. Designers applied the knowledge they had gained from the recently completed Niagara Falls project to the dam in Cornwall. Hydro's lead engineers determined all aspects of the river flow and soil content before putting contracts out for bid.

Keith Henry

After the St. Lawrence project was approved by both governments, the Joint Board of Engineers agreed that all of the models should be built by Ontario Hydro engineers, since we had previously completed the ones for the Niagara Falls project. Don Harkness served as the supervising engineer for the Ontario Hydro Hydraulic Lab. I was designated construction engineer and Bob Johnson was in charge of testing. Once again there was a debate about who was to do the model testing as there had been on the Niagara project, but this time it was not the case of the Canadian mouse and the American elephant. We elephants in Canada had the Niagara model under our belts and the testing had been done so well and so effectively that there was no contest. The U.S. Corps could not muster much support and we had another major job to undertake. For some reason, the Corps thought that the Joint Board of Engineers would award them all of the testing work because they had a more advanced lab in Vicksburg, Mississippi. When Ontario Hydro was selected as the lead agency for this aspect of the project, it was a bitter pill for them to swallow. However, this was an international job being built under the auspices of the International Joint Commission. Commission members saw our models for the Niagara project and thought we were the best men for the job. It was well understood that Ontario Hydro had a tremendous amount of experience in all sorts of power projects from the model to the construction phase.

The St. Lawrence Seaway models and hydraulic studies were conducted in our lab just west of Toronto in the Kipling Transformer Station. The building had been designed for the construction and testing of the Niagara Falls project. The steel building was placed on a concrete slab to prevent the settlement or shifting of our models.

There were two basic types of models. The first was the topographic model showing the current layout of the portion of the river under study. The second was a structural model representing the various dams, channels, and locks to be built by contractors to control the flow of water and improve navigation. We completed these models according to the shape of the land, the river banks, and the bottom drawn by cartographers on large-scale maps. The accuracy of our work depended on the data depicted on these maps and the bathymetric information obtained by echo sounding from a boat.

My colleagues and I built three topographic models. The upstream one or the Iroquois model extended from Prescott, Ontario, and Ogdensburg, New York, to below the Iroquois Dam. The Ogden Island model covered the area from Point Three Points to the Platte Rapids. The Long Sault model extended from above the Long Sault Navigation Channel to below the power house and the Long Sault Dam. We completed these models in less than a year. It was my job to first draw the models on paper and then oversee the construction of the actual floor model. We also built a number of models of various structures, most notably the Long Sault Dam. I was also in charge of the hydraulic model laboratory where we did all the water-level tests for the St. Lawrence. It was quite a hassle. We built five major models encompassing thirty-six miles of the St. Lawrence from the power dam site to the Iroquois Dam.

We also built a number of natural scale models of structures on which to do the detailed hydraulic design. As a starting point, these models were built in accordance with the theoretical preliminary design worked up by the office hydraulic engineers. From then on we modified the model according to our test results. The most elaborate one of these was a 1:60 natural scale model of the Long Sault Dam. Rex Griffiths, an English cabinetmaker and joiner who was our specialist in this work, made a Plexiglas mold of the piers and rollways of the dam which we cast in concrete. That gave us a small replica of each part of the dam with a glassy surface of the correct roughness and dimension. This particular model was used to design the energy dissipation bucket. The proof of the reliability of the structure models for design purposes was the operation of the prototype structures and control mechanisms. To house all of these models we had to extend our existing model building to four times its previous length, but we were able to maintain the same width.

The Ontario Hydro Hydraulic Model Laboratory was, in my opinion, one of the most successful hydraulic design facilities ever built. It accomplished all its tasks in a cost-effective manner and the facilities that were based on its work, some of the largest the world has ever seen, have now been operating for over fifty years. There has never been any criticism that I have heard of any of the way these facilities functioned and I am proud to have had a part in their design. I was then, and still am to this day, extremely proud of being a member of the team that did those major model studies. We worked our butts off to be sure that everything was done as well as we could possibly do it and the results were very rewarding.

Garry Moore

One of the things that I relied on to monitor the water level gauges during the dredging between Cornwall and Beauharnois was the measurements provided by the engineers from the Ontario Hydraulics Lab. A lot of research had gone into the flow of the river. They had constructed these huge models that filled a building the size of an armory that you could walk around in rubber boots. The goal of this research was to determine to what depth each area had to be dredged in order for power production to take place as well as to create a navigable waterway.

Arthur Murphy

Handling the additional water created by the power dam and the increased depth of the waterway was tricky. I remember the original layout from the 1940s that was altered a lot by Ontario Hydro engineers based on their hydraulic testing and structural model. They had the whole Seaway and power dam planned out on paper. Then in their large warehouse, they constructed models that depicted each control structure, including the Long Sault Dam and the Iroquois Dam, and the dredging to determine how all of these facilities were going to function in tandem. They needed to know how all of the dams were going to operate and how they were going to impact the other structures. But a lot of it technically did not work out quite as nice as how they had planned it. Water and natural obstacles like soil and rocks can make the best laid plans and scientific data useless.

The construction of the St. Lawrence Seaway and Power Dam Project took place in several phases. First surveyors and hydrographic engineers arrived to mark the exact location of the various dams and locks as well as determine the dredging levels for dragline operators. Next workers built cofferdams to hold back the water from the dam and lock sites. Once the areas were dried up, earthmovers cleared the required foundation area before the pouring of concrete could begin. Simultaneously, dredging and dike construction were taking place in the surrounding channels.

Robert Hampton

Early in the project my survey group conducted hydrographic surveys at the upstream end of the Long Sault Rapids and in the vicinity of the Iroquois Dam. The areas being surveyed in the rapids sections were between Ogdensburg and Waddington, New York, where the normal river flow was between ten and fifteen miles per hour. The equipment we used had been developed by the Tennessee Valley Authority (TVA). Two representatives from TVA came to Massena and set up the equipment and adapted it for our use. The sonar device was attached to the bottom of our survey boat, which was a twenty-two-foot modified bridge pontoon with a covered area in the stern that was powered by two outboard motors. It recorded the depth of the water as the boat moved out from the shore along a marked range at right angles to the flow of the river. These measurements located the areas that needed to be deepened and determined how much material the dredging contractors would remove at each site. The actual elevation of the surface of the water was established from survey points along the shore. The distance out from the shore was determined by recording the length of a piano wire fastened to a shore anchor that passed over a pulley on the top side of the boat as we went out along the survey line. Once we located and marked each survey point, the wire was cut by a person on shore and the wire was reeled back to its box for use on the next run.

Alfred Mellett

I worked for Uhl, Hall, and Rich doing survey work on the big power dam, the Long Sault Dam, the Iroquois Dam, and all of the channels that had to be cut. The Corps of Engineers designers provided me with the dimensions and location of each structure. The Long Sault Dam was the most difficult facility to survey because it was a curved dam. I laid out the points on the mainland and Barnhart Island where each end of the structure would sit and turned my plans in to the project manager. My measurements were given to the contractors awarded the bid to build the structure. It was then up to them to build it. The location of all of the sites and dredging areas required precision and accuracy, as a mistake on any one of these elements could have led to a costly redo or a ship running aground. My final surveying assignment was plotting the major dredging project on the main channel.

A few months after the project was under way, one of the engineers from Perini, the contractor on the Long Sault Dam project, came to me with a major concern. He questioned my survey measurements because his engineers had found that my estimates were about an inch and three-quarters off. The whole dam was over three thousand feet long, so such a tiny amount made no difference. Regardless, Perini hired a survey company to come in and check all of my work. In the end the curvature was accurate and the dam hit the island. The curve of the structure threw off your visual measurements and if you hadn't worked on water projects before, this concept was hard to understand.

Joe Marmo

While I first began working on the Seaway project, I laid out and did the reduction of the surveyor's notes and the distances regarding the angles on the various locks and dams. The project manager's name was Frank Matayka. I was just a young kid and for some reason we hit it off. One day he came into my office with the surveys for the Long Sault Dam in his hand. He said, "Joe come with me, there is something bothering me." He took me up on the embankment on the mainland to view where the dam was going to be curved to hit an island. He said, "It looks like the way that thing is curving, Joe, it is not going to hit the island." I replied, "You have got to be kidding?" Then I realized he was right! We were both a little out of our element. We had never worked with such a large river, and specifically did not comprehend how water distorted our visual perception. I spent the next week going over every single surveyor's note and measurement I had. When I was finished, I said, "Frank, it has got to hit the island; I can't find anything wrong. I just went over everything." He said, "I hope you are right!" As it turned out the distance of water fooled Frank and me and when the Long Sault Dam was completed it spanned the distance between the mainland and Barnhart Island perfectly.

In the first half of 1954 the employees of Mannix-Raymond Limited of Montreal and Peter Kiewit and Morrison Knudsen of the United States erected three cofferdams, the first elements of the power and navigation project. The large structures dammed up the flow of the St. Lawrence River so that the excavators, ironworkers, and machine operators could build locks and power and spillway dams on a dry riverbed. Initially two cofferdams in the power dam section blocked off the northern channel, and later the south channel of the river. Contractors assembled 12,000 pounds of steel rods into cells of 64 feet in diameter and filled them with 19 cubic yards of soil and rock. Next, workers built a 500-foot-long cofferdam on the north side of the channel. On the south channel from the Canadian mainland to Barnhart Island contractors erected a 4,200-foot-long cofferdam, the largest ever built in the world. Carpenters and concrete crews also labored on the Iroquois and the Long Sault Dam in the "dry." Dump truck operators and welders cofferdammed off one side of the river and then upon completion of the first section of the dam, dredge operators removed the blocking structure and diverted the water through the constructed section.[10]

2. In April 1955 Cofferdam C-1 for the main power dam and the floating bridge from Hawkins Point to Barnhart Island are under construction. The steel-celled cofferdam allowed for the power dam to be completed in a dry area. The bridge allowed contractors to transport equipment and materials from the mainland to the construction site. Courtesy of Alfred Mellett and PASNY.

Donald Rankin

Mannix, a Canadian contractor, built the cofferdams for the power house under the supervision of Ontario Hydro. Upstream there was an earth-filled cofferdam between Sheiks Island and Barnhart Island and a downstream sheet-steel pile-cell cofferdam immediately downstream of the power house. In the summer of 1954, Mannix constructed the upstream earth-filled cofferdam and cleaned up the river bottom for the downstream sheet-pile cofferdam. In the winter of 1954–55, the steel-cell cofferdam was completed. In the spring of 1955 the power house area was dewatered and construction began.

The steel-cell cofferdam had a freeboard thirty feet above the ordinary flow of the St. Lawrence. During the first winter after the power house construction had begun, the winter of 1955–56, because of the blockage of floating ice downstream, the water level rose to a point that reduced the height of the freeboard of the cofferdam to about four feet. I, for one, was certainly alarmed by the great rise of water. I do not know if that particular occurrence worried the people involved in monitoring the river flow. To an onlooker, this was of great concern, because the breaching of the cofferdam would have flooded the work that had already been done on the power house and delayed the schedule significantly. Fortunately, the water receded, leaving the cofferdam uncompromised and the construction schedule intact. However, the Americans were not so lucky. The steel on one of the sixty-five cells on their side of the cofferdam was weak and the cylinder collapsed before construction on the site started. That particular cell had to be removed and replaced.

James Romano

When I worked on the power and Long Sault Dam, contractors put cofferdams across the river to block off the water. Workers pile-drove tubes into the ground and filled them with gravel. These cells were installed alongside each other and then connected. Over sixty cells were placed around half the river like a horseshoe. Then all of the water was pumped out, which created a dry construction area. Once the water was dried up, the contractors put the footings in and started pouring the concrete. When that half of the dam was done, they took the cofferdam apart, opened the gates, and let the water flow through the first side and started building the other half.

When I worked on the Long Sault Dam, it took two days to put the cells up and two more days to fill them. I hauled gravel around the clock until the cofferdam cells were full. It took two thousand yards. I worked from sunup to sundown. The guy who ran the pit had only a little dinky three-quarters-of-a-yard shovel. My truck's dump had a fifteen-yard capacity and it took forever for him to load up my truck. There were usually several trucks ahead of me, so I waited in line for a couple of hours. During that time, I slept in the truck. When the guy in front of me was done, he blew his horn and woke me up.

Joe Marmo

Several once-in-a-lifetime things happened on the St. Lawrence Seaway and Power Project. The flow of the river was the first thing that made this project so much more complicated than any other in the history of the United States. The river itself presented some unique challenges. When they wanted to close a portion of the river so they could build a cofferdam, engineers had first to figure out how to divert the water. They couldn't just go out and drive in pilings with the water flowing at 300,000 cubic feet per second. It was going so swiftly, it just wiped everything out. So the Corps of Engineers dropped concrete tetrahedrons which were bigger than two cars put together up above where they wanted to stop the water. The crane operator had to time this just right. He picked up tetrahedrons on the mainland and dropped them in the middle of the river. They bounced down the river and the contractors were hoping that their pyramid shape would make the concrete boulders stick in the spot where they needed to block off the river. Initially a couple went straight through, so workers went down and picked them up and brought them back to the starting point where the crane operators dropped them again. Eventually they built up a big pile of rocks that diverted the river. The whole process took several days. My friends and I would go out and watch that really unique operation.

Roy Simonds

I think one of the most interesting things that I saw was the dewatering of the Long Sault Rapids. When they blocked off the water, the area was full of large boulders, cannonballs from the American Revolution, and anchors from ships that had landed in the rocks. Some of these large rocks were on top of each other. The current had ground them against each other for such a long period of time that the one on the bottom had a hole in it. That is how powerful that water was.

The cofferdam around the rapids was not made of steel cells; instead crane operators dumped in big boulders to stem the flow. Half of the materials got washed down the river by the rapids, but eventually they made it across. Other areas around the Long Sault Dam, the Iroquois Dam, and the main power house were dried up when contractors built steel-celled cofferdams. The placing of steel pilings was a big job and required large cranes and very skilled equipment operators. Dump truck drivers hauled in gravel from off-site pits to fill these cylinders. These structures were massive and were supposed to keep the construction sites dry. However, during one of the winters there was an ice jam and the water welled up and almost came over the top of the cofferdam at the power dam site.

Ron Barkley

The dams were all constructed on dry land surrounded by cofferdams. The cofferdam at the Long Sault Dam site was a couple of feet wide on the bottom and it had a two-lane road on top. I actually drove out on the dam. It wasn't a flimsy little dam; there was no way in the world it could collapse. It was impossible. Sometimes truck drivers had to position their vehicles right on the edge of the dam and dump their load off the side, but there was no danger of the dam breaking. When one side of the Long Sault Dam was done the cofferdam was taken out and the water was diverted through the gates on the finished half. To remove the cofferdam the workers started in the middle and dragged the bottom with big draglines to remove all the built-up earth and rocks. Then crane operators scooped the gravel and dirt out of the middle of the cylinders. Finally, barge operators removed the empty steel cells by attaching chains to them and dragging them back to shore and gradually the water started coming through.

The Canadian and American engineers and contractors faced several problems that had to be addressed as construction commenced. Firstly, the Corps of Engineers and Ontario Hydro pledged to maintain a fourteen-foot water level in the existing American and Canadian canals to allow shipping to continue during construction. Contractors under their jurisdiction also could not alter the current flow into the Massena power canal that fueled Alcoa's machinery. Secondly, the construction on the various sites, particularly the dredging components, had to adhere to a stringent schedule so as to not adversely impact the work on the other sites.

Many of the design plans remained flexible throughout the construction phase because what looked good on paper or on a smaller-scale hydraulic model did not always translate effectively to the actual construction site or section of the river. Both Ontario Hydro and the Corps of Engineers assigned river flow engineers to Cornwall and Massena to help interpret their models and hydraulic data and to ensure the safe passage of ships on the old canals.

Jim Cotter

As several structures were being built in tandem and boats were still utilizing the old canal system, maintaining the uninterrupted flow of the river was critical. Therefore each project was designed to be built in stages so as not to adversely affect the water levels or flood the other construction sites. Each segment of the Seaway and Power Project began with a dredging component followed by the building of cofferdams. Engineers had to monitor how much earth was being removed in each area to make sure they were in line with their comrades downstream. Once each area was dried, the pouring of concrete and setting of forms began. As the work proceeded on the Barnhart Island Dam and Power Plant, the Long Sault Dam, the Massena Intake, and the Iroquois Dam, the goal was to complete all of the structures and have them operational by Inundation Day. Contractors and high-level officials on both sides of the border had to coordinate their work. American site managers and area district officers met daily to discuss progress on their operations and synchronize their construction plans. They also met with their Canadian counterparts weekly to stay abreast of each others' movements.

Keith Henry

When the actual construction got under way, it became evident that someone was needed in the field who understood the plans for excavation and diversion and the interrelations of the different sites. Each cofferdam, channel, and dike affected the river current and levels miles away from each structure and changed the working conditions at other sites. I had studied these relationships, but it was impossible to know exactly what conditions might arise. We were ready to deal with delays and changes in design due to unforeseen issues.

I really got to understand how the river was changing by talking and riding with a lot of the pilots and the captains of the riverboats as they were going up and down the St. Lawrence. Those fellows were very surprised when we started changing things around the Iroquois Dam site and put the cofferdam out in the river. Some of the older pilots had been navigating that stretch of water all of their lives and all of a sudden the river changed. They didn't know exactly what to make of that. They knew what the river had been like before and they were surprised when currents changed from one place to another. They could read the water pretty well, but they didn't understand why the flow was different. I knew the reason because I had built the models. How it panned out in the river, however, was a little different.

I was on a boat one time when it was approaching old Lock 25 near Iroquois, Ontario. Coming up the river the pilot was guiding the boat. The river currents had changed and I knew how they had changed, but I was standing back. I wasn't telling him a damn thing. I was watching to see what he was going to do when he got into a different situation than he understood. All of a sudden he was headed for the guide wall at the dock. What surprised me was when he just stepped back from the wheel and said, "He is all yours, captain." The captain stepped up and took the wheel. He reversed the engine and did what he could, but we connected with the lock wall hard enough to move the boat about sixty feet. It didn't do the ship much damage, but I can still remember the sound. It was a great big deep bell boom. It was pretty wild water before we stabilized it.

Garry Moore

There was a tremendous amount of cooperation and hydraulic research concerning the configuration of the flows of the river between the Corps of Engineers, Ontario Hydro, and the Canadian Seaway Authority. Many areas of the river had to be dredged to harness the current and accommodate shipping, while other sections remained undisturbed. The International Joint Commission required that the flow on either side of Cornwall Island not be changed from its natural path. In order to do that, contractors dredged the Seaway Channel on the American side between the U.S. mainland and Cornwall Island. Ontario Hydro dredged on the Canadian side to balance the flow of water down both sides. Lake St. Francis, which is the head pond for the Beauharnois Dam, was an area that was heavily dredged and required a lot of surveying and coordination between the engineers and the dredging contractors.

I worked for a year between Cornwall, Ontario, and Beauharnois, Quebec, on the major portion of the dredging on the Canadian side. To guarantee ships safe and unhindered passage, contractors needed to create a required twenty-seven-foot draft, but we tried to err on the cautious side and aimed for thirty feet. It was part of my job to maintain thirty-five water level gauges from Cornwall to St. Aniset. The dredge operators used these water level gauges to judge their depth. They would read them six times a day and adjust their machinery according to the fluctuation in the water level. There was a lot of research that went into the flow of the river. The measurements I followed were based on the Corps and Ontario Hydro hydraulic models in Vicksburg, Virginia, and at the National Research Council in Ottawa.

At the Eisenhower, Snell, and Iroquois Locks as well as at the main power dam site, workers cleared the construction area before the commencement of the pouring of concrete with thirty graders, two hundred tractors, and sixty scrapers that could handle between twelve and twenty yards of material. As glacial till and marine clay covered most of the construction sites and the plans required excavators to dig down to the bedrock, the equipment used in this undertaking had to be more durable than on past projects. The dirt and rocks removed by the operators of this large fleet of earthmovers was loaded into one of three hundred

3. Bulldozer operators are shown leveling the base of the Eisenhower Lock. A Euclid truck is parked in the foreground. Courtesy of Cornelius J. McKenna.

large Euclid trucks with hauling capacities of twenty cubic yards. The Euclid Company of Ohio, which specialized in earthmoving equipment, manufactured these first of their kind fifty-ton, three-axle dump trucks to withstand the weight of the large amounts of material that needed to be transported during large construction projects.[11]

Ted Catanzarite

The equipment the contractors used was fantastic for that time. When I worked as an oiler on the Eisenhower Lock and Iroquois Control Dam, I drove around the Euclids, the big dump trucks with the large tires, and other excavation equipment. Several of us young guys used to oil the bulldozers and bring in the trucks to be repaired by a team of mechanics. One of the things we liked to do was put the blades of two bulldozers together and turn them on. We would push them against each other just to see the ground churn up. One time in the late summer when it was still warm, we had moved the bulldozers to a certain spot on the site to clean them and oil them so they would be ready for the operators the next morning. The operators were excavating a section of the base of the dam and the sun had made a crust on the surface of the mud. I was driving around a bulldozer and I went over some mud that I thought was solid, but it was actually liquid underneath. So, when I drove into the mud, I buried the machine right up to the seat.

My supervisor was an alcoholic and always drunk on the job. He was not a happy drunk. He told me and my buddies that we had to get underneath the radiator and attach a chain to a hook that was meant for towing the bulldozer. So I had to get in the mud and hook up another bulldozer to pull that one out. It actually took three others to pull out the one I got stuck. I could have gotten canned for that incident. Instead I spent the rest of the night cleaning off the machines.

Roy Simonds

The equipment was mammoth in the 1950s. The excavation contractors brought the big shovels in from the gold mines in Kentucky. They had one called the "gentleman" that could pick up a truck in its shovel. I worked on a dam previously and most of the equipment was not quite as big. Equipment was also changing from many of the archaic dinosaurs from World War II to the cranes and trucks that contractors still currently use.

Thomas Rink

The equipment stuck in my mind. It was man-bearing. I can remember those old shovels and bulldozers. They called them man-lickers. Today everything is just so different. Most equipment has cabs, air conditioning, and heat, and the hydraulics are power shift. Back then, the operators shifted by hand. It was a whole different world in terms of machinery. They were all standard shift with no power brakes or steering. It was quite a project. Men were men when they did that job.

Garry Moore

A lot of the equipment and construction techniques utilized and invented by the Seaway workers and contractors was later used all over the world on other waterway and building projects. The tower cranes that steel

4. An earthmover is pushed from behind by a bulldozer during excavation at the Eisenhower Lock. Courtesy of Cornelius J. McKenna.

contractors employ on high-rise buildings across North America in the present day were first used on the St. Catherine Lock. The crane is composed of a great big tower going up in the air with a big boom attached on top. Over the entire five years of the power dam project there were technological improvements that were made in all aspects of construction from concrete, design, and equipment. Even the ten-yard concrete trucks that you see driving around today bringing concrete to various construction sites were a big deal when they were developed in those days.

Once construction got under way, the large gantry cranes and their skilled operators played a key role. The one-hundred-foot-high electrically powered gantry cranes moved along tracks at five miles an hour and could lift a four-yard concrete bucket that weighed eleven tons at the end of their 155-foot booms. Experienced operating engineers sat in the main cab of one of the four cranes on the power dam project lifting concrete forms into places and transporting supplies from one end of the site to the other. These machines were also crucial in the placement of turbines and doors on both sides of the border. Before their installation, several freestanding cranes had collapsed because of their inability to handle the heavy lifting. The forty-four million dollars' worth of machinery used by equipment operators equaled the impressive concrete structures.

The central part of the power project, the jointly built $600-million international power dam, spans 3,230 feet and stands 167 feet high. Five contractors under the supervision of B. Perini and Sons of Massachusetts constructed the U.S. side and a consortium of seven contractors completed the other half under the auspices of Iroquois Constructors Limited in Canada. The two equal sections contained sixteen turbines with a combined annual power generation capacity of 2.2 million horsepower. Each country's contractors constructed a power plant at the beginning of their section that housed transformers that sent power to adjacent transmission lines. Workers built the structure with 1,890,000 cubic yards of concrete, 116 million pounds of reinforced steel, and 15 million pounds of structural steel. The dam engineers harnessed the energy of 81 feet of the natural 92-foot drop of water level between Lake Ontario and Massena. They also created a power pool behind the dam called Lake St. Lawrence to hold the water needed to turn the turbines and generate power. Finally, workers built 18 miles of dikes to contain the power pool.[12] While the Canadian and American contractors employed different

5. On September 17, 1957, the walls of the upstream face of the St. Lawrence Power Dam, which contains ninety-six openings between the United States and Canada, were clearly visible. Courtesy of Alfred Mellett and PASNY.

concrete pouring techniques and construction methods, an amicable working relationship existed from the administration level to the dam floor.

Barbara Brennan Taylor

There were seven companies on the Canadian power project which combined to create Iroquois Constructors. Each of these companies had their own project engineers. The company was put together for this project because there was no one Canadian contractor big enough to handle the building of the power house and the power dam alone. There were all of these top people from all of these firms trying to work together to get the project completed. They were all big shots in their own right, so they all had big egos. I was always aware of the fact that each thought that he was the most experienced or the most knowledgeable or the most educated. They were not easy with each other. The engineers and foremen were certainly all experts in their fields. There was never anything that I recall that was explosive. It was just interesting to see so many people who had run their own show try to run this one together and decide who had the final say. There was also an awful lot of money involved. I think it might have been the biggest project in Canada. To combine seven different construction companies into one to complete the power dam was not easy. I don't know if it had ever been done before or since.

Roy Simonds
Uhl, Hall, and Rich did the complete design and major construction of the American side of the power project. The dam was completed by contractors hired by both the New York Power Authority and Ontario Hydro. I had a lot of interaction with the Canadians, mainly to try and coordinate our work and get ideas from them regarding their problems and try to work out common solutions. Primarily the excavation and concrete curing were challenging and I wanted to know how they were doing with their contractors, as far as the removal of the till and the integrity of the concrete was concerned. I think they were more experienced with cold-weather concrete as far as cost and time. Even with our different techniques and design, each side of the power dam turned out to be the same.

Bart Whitten
The power dam was a massive job and my Ontario Hydro co-workers and I were under a lot of pressure to get all of the various aspects of the power dam built correctly and on schedule. Since it was a joint project with the U.S., we had to meet their deadlines as well as our own. I therefore worked longer hours than on previous jobs. There were men pouring concrete around the clock. There was a lot of pressure grouting under the dam because there were eighteen-inch voids between the bedrock that was ninety feet below sea level. Ontario Hydro contractors needed to consolidate the rock under the dam because of the weight of [the] power house, so there would be no movement. The 1944 earthquake tipped rocks up

6. On December 17, 1957, a 150-ton run assembly for turbine unit 30 is lowered into position on the American side of the power dam. Courtesy of Alfred Mellett and PASNY.

at thirty-degree angles, like a tent roof, and fractured it. Concrete crews pumped in 400,000 bags of cement, sand, and chemicals to solidify the bedrock. This work was time-consuming, expensive, and caused a lot of grief. It was a fractured mess and cost Hydro a lot of money.

Joseph Couture

The Canadian side of the power dam was 156 yards long and we had to meet the Americans halfway. Our biggest concern was that once each country's contractors completed their half, they lined up. When I arrived on the power dam site, Ontario Hydro officials had a firm completion date and a lot of concrete to pour. I was told the structure needed to be finished on time. Therefore we worked in all types of weather and only shut down for two or three days in the winter. Crews poured concrete in high winds and the cold and covered the forms and heated the area. Men were hired just to look after the heating aspect. Ontario Hydro and its contractors were used to these types of projects and weather conditions. We just went out and did it.

I changed the old system of pouring concrete and many of the older supervisors did not like it. Many of my bosses were used to working in the old way. I built a new concrete form in the carpenter's shop that was twenty-five by twenty-five feet and could withstand the pressure of a forty-two-foot pour. It had to be put into place with the cranes that were eighty-five feet off the ground and could lift almost anything. My boss told me I could use the new form if I was willing to risk losing my job if it failed. My new design allowed us to finish two months ahead of schedule, and when I assumed my position with Iroquois Constructors, they were a month behind.

George Haineault

The Robert H. Saunders–Robert Moses Dam was a shared project with the Americans. Both sides had their own concrete mixes and pouring techniques. The Americans had a better system because their panels were smaller. The concrete crews only made five-foot pours compared to our twenty-four-foot ones. The Canadian forms were lifted with gantry cranes because they were too massive to be handled by hand. The Americans were okay with their small pours, but ours were so big they had to leave a space between them until they cured. The dam wall looked like a checkerboard. The crane operators would pull the panel off and laborers would strip it and clean off the excess concrete, and then the next panel would be moved over. I had to watch myself when the new panel came

7. This July 19, 1957, photo, taken from atop one of the gantry cranes, shows the second stage of the Long Sault Dam construction. The final shape of the structure is evident. Courtesy of Alfred Mellett and PASNY.

in because I was on the edge of the dam and if I wasn't paying attention I could get knocked off. It was part of my job to anchor the new panels to the structure below before they were filled with concrete. Then I went back and stood on two twelve-by-twelve-inch beams and filled in the gaps between the pours.

In the main channel between Barnhart Island and the Canadian mainland, Walsh, Perini, Morrison-Knudsen, Peter Kiewit and Sons, and Utah Construction Company constructed a control dam called the Long Sault that closed off the channel near Barnhart Island. Contractors completed the spillway dam exclusively on the American side, three and one-half miles upstream from the power dam. The 2,960-foot-long, 145-foot-high Long Sault, composed of five million pounds of reinforced steel, four million pounds of structural steel, and 660,000 pounds of concrete, controlled the water level in the power pool. The facility also funneled excess water around the power dam into the south channel.[13]

The Iroquois Dam, located twenty-five miles upstream from the Robert Moses–Robert H. Saunders Power Dam, cost $14.4 million dollars and regulates the outflow of the Great Lakes. Engineers classify the facility as a buttressed gravity structure. It consists of a series of piers and sluiceways and has two 320-ton gantry cranes that operate the structure's thirty-two lift gates. Workers poured the first of 194,920 cubic yards of concrete in November 1955 and completed the structure in March 1958. The dam spans the width of the St. Lawrence River from

8. In this May 18, 1957, photo, the first stage of the Iroquois Dam is complete. The construction of the second stage of the dam is under way. The eventual concrete upstream end has been poured and the preliminary steel supports for the lift doors have been pile driven into the bedrock. An unused empty cell of the cofferdam is also visible. Courtesy of Alfred Mellett and PASNY.

the U.S. mainland near Waddington, New York, to Iroquois Point on the Canadian
shore. Peter Kiewit and Sons and Johnson and Johnson constructed the Iroquois
Dam alongside the Canadian Iroquois Lock that allows ships to bypass the dam.[14]

Alfred Mellett

Workers completed the first section of the Long Sault Dam and then the
whole river was channeled through that side and the other half was constructed. The Iroquois Dam was also built half at a time. I placed all of the
markers and outlined all of the specifications for that structure and gave
them to the contractors. The Long Sault Dam was complicated because of
its design and because it was built in several phases. While it was under
construction it was one of my favorite sites to photograph, especially from
the air. It had all of these interesting features including the cofferdam
with the roadway on top, the supply bridge, and eventually the flowing
of the current through the completed side. Contractors also used gantry
cranes to lower the gate doors and for some reason they always did this
at sunrise.

Jim Cotter

The Iroquois had to be a staged project to accommodate the river flow
while each portion was being built, but it wasn't as complex as the Long
Sault Dam, which was built in three stages. The first was the construction
of the cofferdam. The second was the building of the first portion of the
dam. When that was finished, contractors rerouted the river through that
side. The final stage was the completion of the second half of the structure
and the opening of all the gates on Inundation Day, which allowed the
river to flow through both sides of the dam.

Arthur Murphy

The Iroquois Dam controls the elevation of Lake Ontario and the flow of
the Great Lakes. That control dam is a major part of the whole Seaway
system. If there is an influx or major surge of water, the dam operators can
control it right from there. They can regulate the maximum flow down
through the channel so it won't flood Lake St. Lawrence and affect the
power house or the control structure at Long Sault. The Ontario Hydro

engineers did a lot of hydraulic studies and constructed models that illustrated how the whole Seaway and power system, including the dams, was going to operate. But a lot of the components didn't technically function as well as the designers had predicted. Most of the time the gates at the Iroquois Dam are open, but it is there in case of an emergency.

John Bryant

The Iroquois Dam was built by the Power Authority. I remember that there was an argument that it didn't really need to be built. But they weren't sure. If the levels of Lake Ontario started to rise and head down to Montreal, somebody's hide was going to be hung out. The dredging contractors channelized clear up to the Thousand Islands. All through that area, they removed all of the natural control features, so once they opened it up an artificial control structure was needed downstream.

John Dumas

I live in Waddington, New York, which is about five miles from the Iroquois Lock and the control dam. The control dam was one of the biggest boondoggles ever. Ninety percent of the time, the gates on the dam are all open. I can actually drive my boat through the dam and go from Massena to Waddington and on up to the Thousand Islands. If I can drive through the dam, then why did they build it? It has been open for the last fifteen or twenty years. They still do occasionally close it when the water level rises.

Morrison-Knudsen, Peter Kiewit and Sons, and Badgett Mine Stripping Corporation completed the final section of the Seaway project in Massena, the ten-mile Wiley-Dondero ship channel. The bypass waterway allowed ship captains to navigate around the power dam and accommodated two 860-by-80-foot locks—the Snell and the Eisenhower. The channel varied in width, 442 feet at the bottom, 550 feet at the surface, and averaged a navigational depth of 27 feet. The Badgett Mining dredge operators from Kentucky removed four million tons of earth. Workers used the extracted dirt to build dikes around the power pool, the Massena Intake, and along the banks of the channel.

The channel improvements upstream in the Thousand Islands section also created a steady flow to the power pool and removed many natural barriers to eliminate the meandering course of the river. The most extensive dredging involved the removal of fifteen million cubic yards of earth and rock from Galop and surrounding islands. As the work on the island continued twenty-four hours a day, contractors constructed a complete town site that included dormitories and a.cafeteria. The undertaking stood as the second largest earth removal operation in the world. Dredging at Chimney Island farther upstream also improved the navigational aspects of the river. Floating dredge crews completed the majority of the riverbed extraction at this location under water.

In terms of dredging, contractors used a fleet of tugs, barges, supply boats, and floating and standing dredges to create a navigable channel. Because of the inaccessibility of many of these areas in the winter and fall, in the remaining six months contractors scheduled their workers on shifts around the clock. Operating engineers manned ninety-five draglines in the International Rapids section in July 1956 ranging from the huge monighans with fifteen-cubic-foot capacities to smaller shovels with four-cubic-yard capacities. Workers erected the 750-ton walking monighan at the Long Sault channel excavation site the following month. It took twenty-two gondolas and flatcars to transport the dismantled machine from Wyoming to Massena. Peter Kiewit's and Morrison and Knudsen's operating engineers found the monighan's thirteen-cubic-yard bucket and ten-cubic-foot dragline with a two-hundred-foot boom unequalled in terms of removing the shoals in various sections of the river. These operators had to be careful when using this particular piece of equipment because it was only on loan.[15]

Other unique dredging equipment operated on the project included two walking draglines—the "gentleman" and the "junior"—and several large floating barges that also served as lodging for workers. One of the largest dredges, owned and operated by Canadian Dredge and Dock, also included sleeping quarters and a dining area. On the American side, the Badgett Mine Stripping Company tackled the difficult till and marine clay with a 680-ton and a 180-ton dragline called the "gentleman" and the "junior" that had previously been used by strip miners in Kentucky. The 900-horsepower diesel engine powered the 75-foot-tall "gentleman" and its 165-foot boom. The giant bucket attached to end of the dragline, the largest piece of equipment on the project, could dig out a house foundation in two scoops. The "gentleman's" operator dumped the material he extracted from the river bottom into waiting barges purchased from James Hughes and Sons, a

9. The one-hundred-foot Snell Lock walls are close to completion. The checkerboard pattern of the concrete pours and their uniform size are visible. The site also is littered with debris. Courtesy of Alfred Mellett and PASNY.

prominent barge broker. Previously the barges had carried railroad cars from various New York Harbor piers to transfer terminals.[16]

Joe Marmo

One of the biggest sections of the dredging was where almost an entire island was removed. The contractors had a machine called the "gentleman." It was so big, they had to transport it from the coal mines in Kentucky on a series of waterways. You could drive a car into the shovel. The various excavating contractors removed exact amounts of rock and dirt to get the water level to the right point. If they excavated and didn't worry about the river, it would drop the levels below the locks, so all of the dredging operations had to be coordinated. They had to be careful not to flood the construction sites by topping the cofferdams or affect the water levels on the existing navigational route. There were ships still using the old canals and that was a headache.

Frank Reynolds

When I first went up to Massena to work for Morrison-Knudsen, I surveyed a dredging project that was outside of the planned navigational path. I said to my boss, "What in the hell are we doing this for?" He replied that we had to deepen the water in that section of the river to reduce the current and increase the volume to allow for navigation, as well as to supply the power dam. We had to balance the current downriver around Red Mills. Ship pilots could not have traveled through Sparrowbush Point unless that section was dug out and the river tamed. The dredgers took out 6 million yards of dirt. The bid on that job was $.92 a yard. Years ago boat captains steered around Sparrowbush Point because if they got stuck, they had to dump their cargo in the weeds in order to break their vessels free.

James Romano

The excavation contractors kept dredging all year round. The biggest shovel up there, the "gentleman," was an eighteen-yarder. The manawalk shovels were the biggest ones around and had eight-foot-long pads. These monsters had booms that were a couple of hundred feet long. Jim and Jack Maser went bankrupt because they were trying to move the dirt with earthmovers. You could do that in the summer, but in the winter the pads would freeze to the ground. In the winter the contractors used the manawalks. The operating engineers used eight-yard shovels and draglines to dig dikes along the banks of the river to prevent flooding. You could drive a car into the shovel. The size contractors usually used on big construction sites held about three yards of dirt. The operators lined up next to each other and dug the ditch out. When the ditch was finished, I hauled in stone and boulders to shore up the bank so it wouldn't wash away. Once the excavators got the Eisenhower and Snell Lock site all cleaned out, I hauled in gravel and big rocks to line the outside walls before they were backfilled. This prevented the ground from shifting and causing the walls to crack and possibly collapse under the pressure of the water when the locks were filled.

Robert Hampton

The river was constricted east of the marine terminal at Prescott, Ontario, by shallow rapids and a group of small islands known as the Galop Islands. Two contracts were awarded in this section of the river, a dredging contract that was under the supervision of the New York Power Authority, and a dewatered open-cut section in the vicinity of the islands and the rapids that was under the supervision of Ontario Hydro. The dredging

10. In May 1959 the Massena Intake structure held back the water in Lake St. Francis, regulated the flow of water in the Massena Power Canal, and supplied water for the Aluminum Company of America (Alcoa) and the village of Massena. Courtesy of Alfred Mellett and PASNY.

was in an area of glacial till that had been laid down with the melting of the glaciers many centuries ago. The material was made up of boulders, gravel, sand, and fine soils that had been deposited as the glaciers melted. This material required large and powerful equipment to remove it. Many of these contractors had worked on other water projects before; however, one of the things that surprised everyone was the difficulty of working with the till, especially in the wintertime. It hardened up and they had to blast it with dynamite.

Downriver from this dredging contract, large layers of limestone rock were removed, which required dewatering areas of the riverbed and removing large areas of soil and rock by drilling and blasting techniques.

The river flow had to be diverted around these open excavation areas before the work proceeded. As a section of this work was completed, the river was returned to these channels and the opposite side was dewatered for the removal of soil and rock. By this process, the sharp turns in the river were modified and removed, resulting in a more even flow of water through each section. In one of these sections, an entire peninsula was removed to improve the flow of the water and to modify the channel alignment. The channel work in this section was divided between the New York Power Authority and Ontario Hydro. The division was not based upon location of international boundaries, but on a practical division of the work. There was always a practical spirit of cooperation between the two entities.

The bulk of the work of widening and realigning the river channels occurred between Prescott, Ontario, to the west and the Iroquois Control Dam on the east. Downriver from the Iroquois Dam, the river was raised to a total of eighty feet at the completion of the two power plants. With the river at its final height, the channel depths and widths east of Iroquois Dam [were] more than adequate to maintain the river flow and channel widths for power generation and Seaway navigation. The main purpose of the dredging was to maintain the existing flow of water out of Lake Ontario after the raising of the river downstream at the proposed new power plants. The river was to be raised a total of eighty feet at the location of the new power plants, approximately thirty-six miles downstream and east of Ogdensburg, New York. Upstream, the main dredging and open-cut excavation started just east of Ogdensburg and extended eastward about twenty-five miles to Waddington, New York.

At the completion of this work, all navigation channels were checked and approved by the U.S. Army Corps of Engineers and the Canadian Seaway Authority. Some high-level Corps officials came and looked at the Seaway before it opened to make sure we hadn't left any lumps of earth in the navigational channel where ships could run aground and get stuck. These top engineers swept the whole area with their sonar equipment. The head surveyors brought in their pontoon boats and made sure the excavators had made a clear channel. They found several areas where dredgers had to clear out a bit more earth before Inundation Day.

Garry Moore

The dredging was certainly a large effort by an awful lot of men and equipment. Each area that was dredged needed to be surveyed first and then plans drawn up to guide the operators. Some of [the] equipment operators were using more modern dredges, but the owners of the Canadian Dredge and Dock and Marine Industries brought in equipment that they had used since before World War II. They were real monsters, but they really worked well. On one of the largest dredges, the whole crew lived on board.

When the dredging stopped in the winter, my fellow surveyors and I laid out parts of the Seaway channel on Lake St. Francis on the ice. In the spring we gave all of these reference points and ranges to the dredge operators to help them locate the specific areas in the river that needed to be deepened. Any oversight by any member of the surveying or dredging crews could lead to ships running aground once the waterway was opened to traffic.

David Flewelling

At that time the dredging and excavation on the Seaway, particularly at the lock and dam sites, were the largest earthmoving projects in the world. In terms of equipment, it was an evolutionary time as we were going from the steam shovel era to the power shovel era. Contractors were using big scrapers, walking draglines, and gantry cranes that were lifting buckets of concrete that were very state of the art. Some of the earthmoving contractors, however, were from the Midwest and used the same equipment on the Seaway they had used for decades in strip mining, including Perini, who had mined coal in Pennsylvania during World War II. However, even with this experience, the company's owners did not make a profit or even break even on the Seaway because of the marine clay and glacial till. Most of the earthmoving contractors had a fairly good background and knew something of what to expect, including Perini estimators, who had investigated the different materials at each excavation site when they were preparing their bid documents. I just don't think any of them expected to encounter the difficulties and the stubborn material they did.

The remaining supporting facilities built in Cornwall and Massena by Sea-way contractors from 1954 to 1958 included the Massena Intake, the interna-tional bridge, and the relocation of Highway Two. The first undertaking, the Massena Intake, situated five miles from the Eisenhower Lock, involved the reconfiguring of the former Massena power canal to serve as the municipal and industrial water supply. First dredgers removed 2.4 million cubic yards of earth from the area. Then workers erected 180,000 cubic yards of concrete controls that were 740 feet long and 118 feet high. Morrison-Knudsen also con-structed an earthen dike on one side of the new channel to protect the adjoining areas from flooding.

The 3,840-foot-long suspension bridge contained two 232-foot towers sup-ported by two piers, and spanned the St. Lawrence River between Massena and Cornwall. The new structure replaced the Roosevelt International Bridge, which was a combination rail and highway bridge. McNamara Construction Company of Canada completed the foundations for the bridge and Merritt-Chapman and Scott performed the dredge work and the removal of a 367-foot sunken span of an old bridge. The American Bridge division of U.S. Steel Corporation built the superstructure in five months. The St. Lawrence Seaway Authority of Canada and the St. Lawrence Development Corporation on the American side divided the cost of construction based on the amount of miles of bridge in each country. The new bridge was open, to car traffic only, on December 1, 1958, as the New York Central Railroad canceled their passenger and freight service between the United States and Cornwall. The final segment of the power and Seaway construction on the Cornwall side was the 7.2-mile diversion of Highway Two around the Ontario Hydro power house at a cost of $435,220.

The flooding of 22,000 acres along a thirty-five-mile stretch of the St. Law-rence River on the Canadian side of the shoreline remains the most controver-sial part of the St. Lawrence Seaway and Power Project. The residents of the towns of Iroquois, Aultsville, Farrans Point, Dickinson's Landing, Wales, Mouli-nette, Milles Roches, and part of Morrisburg, Ontario—known as the "Lost Villages"—either sold their homes to Ontario Hydro or had them moved to a new lot. Ontario Hydro officials created two new communities, Ingleside and Long Sault, and constructed 6,500 new homes, two shopping centers (including a $1.5-million-dollar strip mall in Morrisburg), six new schools, two community centers, and nine churches. Officially house movers transported the first house *on August 2, 1955, and the last on December 5, 1957. Experts estimated that the actual cost to move a house averaged between $3,000 and $11,000.[17] William J. Hartshorne, a house mover from New Jersey, and his twelve employees transported a total of 531 houses on his one-of-a-kind electric house float that lifted the houses off their foundations by placing two metal prongs through holes drilled in the basement walls. Contractors also relocated the headstones or remains from eighteen area cemeteries and covered the other grave sites with heavy stones so the caskets would not wash away during the flooding.*

Les Cruickshank

Ontario Hydro constructed two new town sites and hired Bill Hartshorne from New Jersey to move the houses from the towns that were going to be flooded to their new locations. Originally all of the old houses were going to be demolished and new homes constructed. However, Bob Saunders, who was the chairman of Ontario Hydro and the former mayor of Toronto, decided it would be cheaper just to move some of the houses because they were in good shape. Somewhere along the way he heard about William Hartshorne. Hartshorne was probably in his sixties then and he had an eight-millimeter promotional film showing his innovative house-moving machine. Bob Saunders thought it was the most incredible invention he had ever seen. Old Bill came up here with fifteen men and his equipment and moved all the houses for Ontario Hydro. At the time, homeowners had three options, if their house was moveable. Hartshorne would move it and put it on a new lot with a new basement, electricity, indoor plumbing, central heating, and town water and sewer. The second option was to receive a cash settlement. The final option was to be given a lot in one of the new townships and to build their own house on.

Ontario Hydro employed property guys to make purchase deals with homeowners. Some elderly couples weren't up to speed with what was going on financially, so they got shortchanged. The majority of the people, in my opinion, don't have much to complain about. Ontario Hydro rebuilt the roads, the churches, and the schools, and gave us new shopping centers. Many residents of the Lost Villages got new homes that were cheaper to maintain and more valuable than their old ones.

Glenn Dafoe

The actual moving of the houses didn't start until 1955. Initially, at the new town sites, Ontario Hydro workers installed water and sewer lines and electricity. They also conducted surveys and laid out lot plans. When we arrived to prepare the land for the new towns, it was just open fields and pastures. Once the surveyors staked out the plot plans, we started the excavation for the foundations and the footings and the masons came in and put up the walls.

The moving process was quite a feat because there was no need to remove anything in the house that was being relocated because of the way the mover was constructed. This was the brainchild of Hartshorne himself. He adapted and expanded on the motorized part of big earthmovers known as Latourneau Scrapers. The house mover had a motor out in front and huge ten-foot wheels behind. There was also a series of cables to hold the house in place. Before Hartshorne moved a house, surveyors took measurements because a lot of houses were old and weren't sitting perpendicular or perfectly flush on their foundations. So as not to jeopardize the structural soundness of the homes, we attempted to place each one on its new foundation exactly as it sat on its original one. Once Ontario Hydro officials decided that a house was going to be moved, workers went in with jackhammers and put holes in the stone foundations. Then men skidded large I-beams through the holes in the walls. The driver of the house-moving vehicle backed in, attached cables to these beams, raised the house off the original foundation, and onto the bed of the float. It is probably primitive by today's standards, but it was a marvelous piece of equipment back in those days.

An elderly couple was the first to sell their home in Aultsville, Ontario, to Ontario Hydro. It was a two-story frame home with three bedrooms, a large kitchen, dining room, and parlor, and it was in very good condition. Ontario Hydro property agents offered them $6,000 for their home and they immediately accepted the purchase price. That sounds rather ridiculous today, but in all fairness if the Seaway had not gone through and that home had gone up for sale on the regular market it probably would have sold for about $1,200. However, another homeowner who was a little more hard-nosed might have argued that since the project was leaving the area with a Toronto-level cost of living, $17,000 was a more reasonable price for a similar property. It raised the price of real estate around this area. If people had banded together and shared information about what Ontario Hydro was offering them, everyone would have benefited more financially.

An Aultsville widow and her daughter were very salt-of-the-earth people. She was intimidated by all of the uproar and was worried that she was not going to have a home. Ontario Hydro representatives made her a purchase offer which she eventually accepted. On the day she signed her purchase contract, a representative from a company in Hamilton, Ontario, came in and said he could build her a new home for the exact sale price of her old home that would be maintenance free. She signed an agreement with him and gave him every cent she got from Ontario Hydro. She didn't realize that her new home was going to be more costly to maintain because of the tax structure and all of the utility bills. There was always someone looking for a quick buck. Residential contractors were in short supply for people who had elected to sell their old home to Hydro and build a new one.

Several cemeteries, including the St. Lawrence Valley Union Cemetery, which was located between Long Sault and Ingleside, were flooded. Plot holders had the opportunity of having their relatives' remains moved or just the stones. Any cemetery that was totally flooded was covered with crushed stone to prevent erosion. The cemetery in Aultsville which actually never was covered with water, was one where all the stones were moved. Not every family wanted the remains relocated. The cold hard facts were that all that were left to move were bones.

John Moss

I moved to Cornwall during the Seaway construction because we were flooded out. I used to live in Moulinette. I was just a renter at that time, so Ontario Hydro gave me the equivalent of a few months' rent. There were a lot of different deals made with homeowners to either buy their homes or relocate them to new lots. There was an army of property men from Hydro and they struck out to make the best deals. They were pretty good horse traders. These men were careful because they didn't want to set an example of paying too high a price, but they had to buy everything in the area that needed to be flooded. Basically I think that Hydro paid pretty well because they could have just kicked the homeowners out and paid them nothing if they wouldn't agree to sell. A lot of area residents complain that they were not fairly compensated for their property, but that is just crybaby stuff. My father was in business here and my grandfather before that and I knew the price of what things were selling for at that particular time and Hydro officials didn't bust their asses giving you big dollars, but they paid reasonably well.

William Rutley

Ontario Hydro had a warehouse where I kept track of all the houses being moved or purchased and the value of each of the dwellings. I made a list of all of the furnishing in the houses the property agents bought and their value so when these items were sold at auction, they made a profit. Every house and building in each of the Lost Villages was assigned a number, so Ontario Hydro officials could keep track of how many houses had been sold or moved to the new town sites. There were also quite a lot of houses that the owners chose not to move, and Ontario Hydro sold them to other people who in turn moved them. It was quite a complicated process, but we developed a simple way to track our progress and manage our accounts receivable.

I found out later that Ontario Hydro burned quite a few of the houses and buildings that they purchased because they were in a hurry. I thought a lot of them could have been moved by private people. Some of them were excelerated [*sic*] with oil and others were just burned by lighting a match. Researchers timed how long it took a house or curtains to ignite and burn. All kinds of buildings and barns were part of the program. A series of nine controlled burns was conducted by the National Research Council and Ontario Hydro involving six homes, a two-story school, and a large brick hall formerly used by the Masons. All of the buildings were different sizes and constructed of varying building materials in an effort to test their resistance to fire. All of the properties were located in Aultsville, an area being flooded as part of the power project. The fire marshall said his main interest was to find out what happened in a burning house in the first twenty minutes of a blaze. The actual findings resulted in changes in the Canadian fire code.

Ontario Hydro oversaw the relocation of bodies and headstones from all the cemeteries that were scheduled to be flooded. Workers moved most of the headstones, but not a lot of the bodies. People didn't want their ancestors disturbed and they are now under water. There was a huge cemetery with a lot of old flat limestone headstones. Those early stones are now displayed on a wall at Upper Canada Village. Ontario Hydro employees tried to locate living members of the family. All of my wife's relatives were buried in old #2, a Presbyterian cemetery at Woodlands. They elected to just have the stones moved. There were cases where people requested that the whole plot be relocated.

I had a first cousin who dug up the graves. I wouldn't have performed that task no matter how much money Ontario Hydro paid me. I was afraid of what I might encounter, including diseases. There were never any problems that I know of where people contracted any illnesses from the corpses, but it wasn't a job for me.

Arthur Murphy

I remember conducting a tour of the project sites about a mile west of Cornwall. Howard Smith Paper Mill, now Domtar, had a whistle that went off every day at noon and could be heard all over the valley. When I arrived at one of the cemeteries where Ontario Hydro grave diggers were removing somebody's remains, I explained to my group how Ontario Hydro was trying to contact relatives whose ancestors were buried there to determine whether people wanted the bodies moved to a new plot or just the headstones. People could request to have their family member's remains moved to a new community cemetery, where the worshippers of different religions were buried. If we couldn't contact any living members of the deceased's family, the grave sites were covered with heavy rocks. As soon as the whistle blew the workers actually stopped, grabbed their lunch pail, and sat by the edge of the grave with their feet dangling, and had their lunch. It was odd to see such a normal activity taking place in such a nonchalant manner in such an eerie setting. These grave diggers were paid a dollar an hour higher than the current laborer's rate to take on this grim but necessary duty.

3

The Workforce

WORKERS WHO PARTICIPATED in the greatest construction show on earth arrived in Massena with different cultural, educational, religious, and professional backgrounds. While the engineers tended to be college graduates, many of the carpenters and laborers had not finished high school. Some stayed for the duration of the project, while others labored for only a few years and then moved on to another construction job. Uhl, Hall, and Rich and Ontario Hydro supplied housing for their administrators and high-level engineers who relocated to Massena and Cornwall with their families, which enticed them to remain in the area the longest. Furthermore, substantial pay raises solidified their commitment to the project until its completion. The other long-term employees of various contractors on the Seaway and dam sites included local unskilled and skilled workers who were established in the community and had no desire to leave the area.

The engineers employed by Uhl, Hall, and Rich arrived in Massena with several years of experience. The Power Authority of the State of New York had hired Uhl, Hall, and Rich to supervise the design and construction of the American half of the Barnhart Island Power Dam, the Long Sault Dam, the Iroquois Dam, and the Massena Intake. Charles T. Main, Incorporated, of Boston, Massachusetts, created the partnership of Uhl, Hall, and Rich to satisfy New York State's engineering laws. For six decades before the Seaway construction commenced, Main had designed and supervised the construction of hydroelectric plants all over the world and employed the most experienced engineers.

The three lead engineers in this new partnership oversaw a design staff in their central office in Boston as well as one hundred engineers and ninety support personnel in Massena. These men had spent their entire careers conducting field studies, and designing and constructing water development projects. William F. Uhl earned a doctorate degree in engineering from Tufts College. His professional credits included serving as a fellow of the American Institute of Electrical Engineers and as a consultant on numerous large dam and hydroelectric projects. As chairman of the board of consultants for the Tennessee Valley Authority (TVA), he assisted in designing and constructing many of the key facilities in the Tennessee Valley. Wilfred M. Hall headed the construction unit on the Wheeler Dam for the Tennessee Valley Authority and oversaw water resource development in Puerto Rico for a decade. George R. Rich earned a doctorate in engineering from Worcester Polytechnic Institute and previously held positions at numerous engineering firms including Stone and Webster and the Corps of Engineers. From 1936 to 1945 he served as the chief design engineer for the TVA.[1]

The office manager and members of the contractor claims team for Uhl, Hall, and Rich had held similar positions on the Boulder Dam and other navigational and power projects under the direction of the Bureau of Reclamation. These men shared their expertise with fellow engineers and skilled workers on the Seaway and Power Project. Some of these individuals served as mentors to entry-level engineers who had recently graduated from college. While being immersed in the design and construction of their first public works project may have been overwhelming at times, the time spent on the Seaway and power dam sites served as an opportunity to put into practice many of the theories that they had learned in the classroom. In the future, this on-the-job experience would lead to their employment by contractors on other large construction projects. These men combined their talents to complete the most complex power dam ever attempted, not only because of its size, but because of the need to coordinate their efforts with their counterparts across the border. They overcame their professional differences to complete the dam on time and under budget.

The administrators of the Bureau of Reclamation, an entity established by the United States under the Reclamation Act of 1902, used funds obtained by the U.S. Secretary of the Interior from the sale of public land

in sixteen arid western states to investigate and construct federal irrigation systems to store and supply water to settlers and farmers. The Bureau
completed 476 dams and 348 reservoirs, and various supporting tunnels,
roads, bridges, and telephone and transmission lines. The Hoover Dam in
Nevada and the Grand Coulee Dam on the Columbia River in the state of
Washington stand as the most famous projects funded by the reclamation
program. After the agency's engineers constructed all of the necessary
facilities, the Bureau became the largest water wholesaler in the United
States and presently operates fifty-eight power plants.[2]

Many of the Bureau men who relocated to Massena had completed
the 564-foot-long Hungry Horse Dam, the highest dam in Montana,
between 1948 and 1953. The facility managed rising water levels caused
by the yearly spring floods of the Columbia and Flathead Rivers and provided electricity to the residents and businesses in the Pacific Northwest
that suffered shortages during the winter months. Besides building a dam
and reservoir, contractors constructed appurtenant works, a power plant,
and a switch yard. The entire structure contained 3,086,200 cubic yards
of concrete. Like the Seaway, it took supporters of this project several
decades to gain congressional approval. Surveyors conducted the original
geological and drainage studies in 1921; however, federal policy makers
did not authorize the funding for the entire project until 1944. The Bureau
and their contractors constructed the dam and its supporting facilities
between 1948 and 1953.[3] Many of the engineers and skilled tradesmen
left Montana after the dam's completion and were immediately hired by
Seaway contractors based on their expertise.

The men who worked on the Hungry Horse Project offered Seaway
contractors a unique knowledge of several key areas of public works construction. Given that all had previously worked on a project that had faced
public and political opposition, they understood the need to stay within
a budget and to adhere to a stringent time schedule to prevent further
antagonism. Also, many had worked in the Bureau's claims department
at Hungry Horse and had dealt with outside contractors and cost overages
as the agency had always put their projects out for private bid. Incomplete plans and soil analysis and change orders did not faze them. Bureau

employees had been exposed to these issues on the Hungry Horse project and when similar problems arose during the Seaway construction, they skillfully evaluated the change orders and sought advisement on the treatment of unforeseen soil and dredging issues.

Joe Marmo

I had the longest tenure of any person on the St. Lawrence Project. I arrived in 1953 and stayed until 1960. I am from Boston originally, but after World War II I played hockey and baseball for the University of Massachusetts at Fort Devans, and then in addition to my GI bill, I got an athletic scholarship to play hockey and baseball for the University of Michigan. Michigan, at that time, was one of the few universities in the country that had counseling to determine what career you were best suited for. When I met with the counselor, I asked him what field I could specialize in that guaranteed me the best possibility of a job. I was a Depression baby, so finding a job with long-term financial security was important to me. He replied that the most marketable degree was civil engineering because the government always needed to build sanitation facilities, highways, and dams.

I graduated in 1951 with a specialty in dam design and the Bureau of Reclamation immediately hired me. However, a year later when President Eisenhower got elected, he cut the staff of all government agencies by 10 percent. Even though my job was protected because I was a World War II veteran, others weren't who had been working for the Bureau for fifteen years. I didn't want to work for the agency any more, so I started looking around. A friend of mine in Massena, New York, contacted me about a position on the St. Lawrence Seaway and Power Project. I sent a resume to the Power Authority office manager, who hired me in early 1954. I had two positions on the Seaway Project: office engineer for the Power Authority on the Long Sault Dam and member of the contractor claims team for Uhl, Hall, and Rich.

Robert Hampton

I am from upstate New York. I graduated from Union College in Schenectady, New York, with a degree in civil engineering in 1945. After graduation, I accepted a short-term position in the Maintenance of Way Department of the New York Central Railroad. I staked out the land for the realignment of mainline track and inspection of railway bridges. Next I surveyed and laid out new routes for electric power lines in the rural areas of the service region for the Mohawk Hudson Power Company in Albany, New York. Following this, I accepted a position with the New York State Department of Public Works. In that position I surveyed and designed sections of the New York Thruway. I also acted as inspector of construction on a section of the New York Thruway in the upper Hudson River and the Albany areas. With the completion of this work, I applied for a surveyor position with the firm of Uhl, Hall, and Rich. I was thirty-one years old. I was brought up to Massena for an interview and was hired soon after. At the St. Lawrence project, I conducted hydrographic surveys and supervised contractors on the upstream channel enlargement and dredging sites.

11. Robert Hampton and his crew prepare to launch their work boat, obtained from the TVA, to undertake hydrographic surveys. Their surveys mapped the bottom of the river and located soil to be removed by dredging contractors. Courtesy of Robert Hampton.

Jim Cotter

I had several years of experience on hydroelectric projects with the Bureau of Reclamation in the western United States, namely the Davis Dam on the Colorado River and the Hungry Horse Dam in western Montana. While the demand for electrical power increased, the sites for hydroelectric developments dwindled. But even though the St. Lawrence Seaway and Power Project stands as the last major dam and lock development in the United States and Canada, many refer to it as the "granddaddy" project of them all. When the Corps and Power Authority contractors accepted the contractors' bids, a great demand occurred among these companies to locate and hire anyone who had previously worked on hydroelectric and waterway projects.

Immediately after his inauguration, President Eisenhower reduced the size of the Bureau of Reclamation. The department employed about 15,000 men. All of the major hydroelectric projects had been completed. I resigned from the Bureau in 1953 and went to work as an insurance adjuster. One Saturday I met a former co-worker of mine on the street who informed me that he was working in Massena on the preliminary stages of the Seaway project for our former boss at Hungry Horse, Chuck Balmetier. Within forty-eight hours I had sent a telegram to Chuck informing him of my interest in relocating to Massena and leaving my boring insurance adjusting job. It was quite a departure and less exciting than heavy construction. The Uhl, Hall, and Rich employment office hired me to deal with contract administration. I arrived in Massena, New York, with my family in July 1955. I spent the next four years on the project.

Jack Bryant

I was born in Colorado in 1926. I worked on the Hungry Horse Dam, a hydro project in northwest Montana for the Bureau of Reclamation. After Hungry Horse I handled construction claims in the chief engineer's office in Denver. The Bureau had done a lot of hydrodams, but it was starting to slow down. All of the main sites had been developed and it was a challenge to get money from Congress for new projects. When this opportunity (Massena) came along, I grabbed it. Then of course the Bureau had a total disaster several years later when the Titon Dam in eastern Idaho, an earth-filled dam near the Wyoming border, failed and flooded the surrounding areas and killed some people. The Bureau, wh[ich] had a top-of-the-line worldwide reputation for dam construction, all of a sudden died.

I heard about the job on the Seaway through friends from the Bureau of Reclamation. Quite a number of the supervisors and skilled workers on the Seaway had transferred from Hungry Horse. For example, the project manager for the contractor at Hungry Horse was also the superintendent for the contractor at the Long Sault Dam. I started working for Uhl, Hall, and Rich in 1954 and left in 1958. I was in charge of the claims adjustment department of the contract administration office. I fought with the contractors over design changes and extra costs. My group of five engineers included Joe Marmo, Roy Simonds, and Phil Wagner.

Roy Simonds

As an employee of the consulting engineers Uhl, Hall, and Rich, for the St. Lawrence Power Project from 1956 through 1960, I worked primarily on contract administration and claims adjustments. My title was chief of contracts. I had worked with the Bureau of Reclamation on a power dam project in South Dakota and most of the engineers on the project had gone from the Bureau of Reclamation to Uhl, Hall, and Rich because of their dam-building experience. I guess it was kind of a domino effect: one person got a job on the Seaway project, and he passed on the information to the rest of us.

The five years I was up there, I was very happy. I look back at it as the best time of my career. Every day was different. Things were always changing and I never got to the bottom of my "in" basket. All of these problems kept cropping up that the contractors would bring to me and expect me to figure out.

Floyd Grant

I worked as a security guard for Uhl, Hall, and Rich, the main contractor, but I was paid by Morrison-Knudsen. I was sworn in as a sheriff deputy. I previously worked for Alcoa as a security guard before the big layoff. I had just lost my job when I heard about the Seaway contractors searching for security guards. When I went to apply for a job, I discovered that one of my bosses from Alcoa was serving as the director of the security force. I didn't even have to fill out an application. He just hired me. I also saw a lot of electricians and plumbers at the employment office who had also lost their jobs at Alcoa. I liked my job because it paid well and it was interesting. I enjoyed being out in my patrol car alone, especially at night.

Before beginning the redesign of the Seaway plans, the Corps of Engineers' personnel officers struggled to find experienced engineers and talented recent college graduates willing to accept their employment offers. The Corps' status as a government agency prevented the Buffalo District Office administrators from raising their approved compensation rates or offering profit-sharing packages comparable to those of company recruiters in the private sector. Prospective employees found every part of the job acceptable, including the challenging work of designing dams, bridges, and locks. However, the salary barely kept pace with the increasing cost of living. For the Seaway project, the Civil Service Commission authorized the Corps to conduct a two-year training program for new junior engineers and offer higher salaries along with benefits and salary increases after six months. It allowed the Corps to hire men with scientific knowledge not specifically in the area of engineering. In Buffalo, college graduates from the Great Lakes region joined twelve engineers from Watertown, New York, and Cleveland, Ohio, to initially work on redesigning the plans and eventually supervise their implementation on-site in Massena.[4]

Kenneth Hallock

I am originally from Marathon, New York. In 1949, I began working for the Corps of Engineers in Buffalo as a junior engineer. Four years later we were charged with a major overhaul of the 1941 Seaway report, the sketchy prewar design. To complete the new plans for the Seaway and Power Project, Corps officials hired many new engineers and office staff. When I got transferred to Massena with five other Corps engineers, I became the chief of structural inspection. My main responsibilities included conducting all of the engineering, mechanical, electrical, and structural inspections.

Most of my fellow engineers hailed from around the country and had worked on other similar projects. Tom Airis, the area engineer for the Corps, had most recently supervised hydro projects overseas, but began his career in the Detroit District. As an ex-military man, he ran a tight and orderly civilian organization. The engineer on the Eisenhower Lock, Dick Wolf, came from a project in the Dakotas. The Grasse River engineer had previously been assigned to the Omaha District. Most of them had overseen other Corps projects around the country and received transfers to the Seaway project when those jobs ended. One hundred and twenty-five Corps of Engineers employees designed and supervised the construction of the two locks, the miles of dikes in between, and the dredging in the International Rapids section.

Ambrose Andre

As a student at Clarkson University, I went on a day trip to Massena with my engineering class to view the initial stages of the Seaway construction. I saw machine operators excavating for the Eisenhower Lock with earthmoving equipment. My classmates and I didn't know much about excavation or how the finished locks and the dams would function. Six months later the Corps set up interviews at Clarkson University to hire civil engineers. I signed up for an interview and went down and talked to their recruiters. They told me about the Seaway and the salary for the various positions. When I returned to my dorm room, I enthusiastically told one of my friends, Robert Carpenter, who hadn't gone to the interview, about the Corps' interesting positions at their district offices in New York. Bob walked down and got an on-the-spot interview and the two of us ended up working for the Corps right after graduation. We went to Buffalo and worked on the lock designs for six months and then we both went back up to Massena in July 1955. Many of the engineers, like myself, were recent college graduates who were working on their first big construction project. Initially I worked in the field office and then as a grouting and concrete inspector on the Eisenhower Lock project.

Robert Carpenter

I grew up near Carthage, New York. I left high school in the middle of my senior year and went to Clarkson as a freshman at the age of sixteen. Nine years later I went to work for the Corps of Engineers. When I went to Buffalo out of college there was a chief of engineers named Stanley Hunt. He singled me out and made sure I got a good start. Hunt was finishing a job he started twenty-two years earlier in 1932 when the Corps prepared the initial navigational surveys when the passage of the Seaway bill seemed imminent. Hunt's aide, Ralph Callinger, had supervised the building of the Mount Morris Dam.

Initially, when the construction began in Massena, I was one of three shift engineers on the Robinson Bay Lock (aka the Eisenhower Lock) along with Nick Patterson and Bob O'Neill. There had to be one of us on site at all times, so we switched shifts every few weeks. If any problems or questions concerning the construction plans came up, I had to deal with it. Eventually I received a promotion to the position of construction management engineer and made sure the contractors followed approved procedures. On the Seaway, the Corps strictly served as an inspection agency. Even though our engineers had designed the entire project, our sole responsibility was to make sure the contractors followed the contractual specifications. Toward the end of the project, the Corps officials transferred me to the Barnhart Island Power House as a resident engineer.

Joseph Foley

I grew up in Fulton, New York, and graduated from Clarkson University in Potsdam, New York, in 1952. Upon graduation, I accepted a job in California working for an aircraft company for three years because they paid the most money. I wasn't doing any civil engineering work at all, and I saw in the Clarkson alumni paper that the Corps was hiring men to work on the Seaway. So I applied and they offered me a job as a low-grade civil engineer. I started in 1955 and I spent my entire career with them exclusively in the Buffalo office. I made several trips to Massena during the Seaway construction to inspect the concrete work on both locks. I spent the remainder of my work life trying to repair the inadequate concrete on the Eisenhower Lock.

Bill Spriggs

I grew up in Ogdensburg, New York. I was a high school dropout and served in the armed forces during World War II. When I returned to the area, I went back to high school and eventually graduated from Canton ATC with a degree in electrical technology. I was hired at Alcoa, but my boss quit, so I left and opened my own electrical contracting business. The assistant area engineer from the Corps of Engineers, Forest Brown, "Brownie," bought a home in Massena in 1952 and I did some electrical and heating work for him. I met him at the Congregational Church where we were both members. He asked me to come and work for him on the Seaway project as an electrical engineer. All I needed to do was to go to the Corps employment office and fill out an application. I went the next day and completed the necessary forms and I never heard anything. A few months later his wife called me to repair her television. She told me when I arrived that Brownie was angry with me because I had never come in to inquire about the position he had offered me. I told her that I had filled out an application months ago. She called her husband immediately and he talked to his personnel officer, who found my application in his drawer. I had supplied my name, rank, and serial number, but no contact information. Brownie asked me to come down to his office immediately. That was on Friday and I went to work for the Corps of Engineers on Monday as an electrical inspector with no previous experience. I was evaluated by a couple of electrical engineers, but I was my own boss. I was twenty-five when I started working there.

Initially in 1956 the excavation was just getting under way and my job at that time was to make sure that the electrical supply lines were safe for the contractors. I worked on the Eisenhower and Snell Locks and on the power plant that they set up along the river to service the areas that were being dredged. I also worked in the Corps office doing lift drawings. The locks themselves were built in forty-foot-wide blocks called "monolithic." Each block had to be inspected to make sure the conduit and piping matched up to the block next door. So the Corps supplied the lift drawings to the various contractors.

I worked with a lot of military types who had spent their careers with the Corps and had experience on hydro projects. However, there were many other positions held by men who were not trained in that particular field. Most of us had no previous experience or skill to do the work we were doing, so we received on-the-job training.

The Corps brought in a couple of seasoned guys when we were ready to open the locks. One had worked for forty years in Panama. He had moved there to play professional baseball and stayed on as the chief of lock operations on the Panama Canal. He was hired by the Seaway Authority to oversee the operation of the locks. When he arrived, he told me he didn't know anything about running a lock, so I should do it. After the completion of the Seaway and Power Project I kept the Snell and Eisenhower Locks in operational condition. On Inundation Day I made sure the doors opened and closed and everything ran smoothly. A few bugs needed to be worked out, but they functioned.

The typical skilled Seaway worker could be characterized as a married twenty-five-year-old man with two children who had previously worked on a hydrodam transmission line project for the TVA, the Bureau of Reclamation, or Ontario Hydro. These men hailed from various regions of Canada and the United States and had earned a living as transient workers, moving from one large construction project location to another. Each had either been trained to operate expensive equipment by an operator on another dam or waterway project or had completed an apprenticeship in their trade. Many of the contractors had a core group of skilled workers whom they brought from one project to another. These men trained recently hired masons, electricians, and carpenters, and based on their past experience received higher pay than their subordinates.

The link between the St. Lawrence Seaway and the TVA began with the inauguration of Franklin Roosevelt as governor of New York in 1928. Roosevelt realized that he had to address the high prices regional power companies charged New York State energy users compared with their Canadian counterparts across the St. Lawrence River. He speculated that state-built dams and power plants on the St. Lawrence would solve this problem, if he prevented the owners of private utilities from accessing the new output unless they lowered their rates. While the fight for public power and the St. Lawrence Seaway initially failed in New York, it steered Roosevelt toward another large-scale project in the Tennessee Valley.⁵

Senator George Norris of Nebraska, an advocate of flood control and power generation in the Tennessee Valley, had supported Roosevelt's Seaway and power proposal. Norris convinced Roosevelt of the feasibility of his electricity proposition, which entailed having a comprehensive government authority that would eventually regulate all of the nation's power generation facilities, but would initially concentrate on completing the dams and waterways in the Tennessee Valley. After the

1932 election, FDR accompanied Norris on trips to Muscle Shoals and the Wilson Dam in Alabama, where the federal government had begun building facilities during World War I to power nitrate production plants. Based on the success of these endeavors, Roosevelt created the TVA as part of the New Deal in May 1933 to complete the navigational and power production facilities in the region.[6]

As the inaugural chairman of the TVA from 1933 to 1941, David Lilienthal spurred the building of twelve dams. Lilienthal touted his ultimate goal as the creation of a system that could supply power to the war industry in the Tennessee Valley and also provide flood control and a navigable waterway for recreational and commercial use. When construction on the TVA project reached its peak in 1942, engineers oversaw the construction of twelve hydroelectric plants simultaneously, with the support of the agency's 28,000 employees. The entire system completed in 1945 encompassed 800 miles of commercial and 375 miles of recreational channels, nine main dams, and four auxiliary locks. Currently the TVA, the Corps of Engineers, and the U.S. Coast Guard jointly operate and maintain the facilities.[7]

TVA planners and engineers had designed numerous facilities to serve both navigational and power needs, and the knowledge they gained from these projects made them desirable employees of the Corps of Engineers on the Seaway sites. During the TVA efforts dredgers completed traversable waterways and harnessed the power of swift currents in rivers and lakes filled with shoals and rapids that historically had hampered navigation. The contractors and designers of these systems had to deal with natural obstacles and the coordination of dredging, lock, and dam construction. The TVA engineers also possessed the knowledge to operate the latest sonar technology in order to obtain accurate depth measurements. When many of the former TVA engineers arrived in Massena in 1954, they had previously dealt with multiple agencies and job sites and a large workforce. Having participated in the construction of a complicated large public works project made them invaluable to contractors and workers who completed the St. Lawrence Seaway and Power Project.

Frank Reynolds

I was a junior in college in 1941 and the U.S. Selective Service was drafting for the war, so I decided not to go back. I saw an advertisement in the paper that the Corps of Engineers was looking for surveyors. I didn't have any experience, but I knew a lot of math. So I got a job as a rod man on a surveying crew for the Corps of Engineers laying out new power lines from

Taylorsville, New York, to Massena. The guy who did the recording on the job died from pneumonia, so my crew chief asked me if I would take his place. I accepted this new role even though I didn't know anything about it. But learning the duties of that new position got me interested in recording and running the instruments and I just fell in love with that work. When I worked for the Corps on the power relocation plan, I was given a six-month deferment. As soon as I finished the job I could have transferred to Puerto Rico for a project and I should have said "yes" because the next week I was drafted into the service. I was part of a ranger unit in World War II.

After the war the Department of Defense hired me to do computer and surveying work. The project manager on the Seaway for Morrison-Knudsen telephoned me in 1955 and offered me a surveying job. My wife, Muriel, and I had just bought a new home. But my elderly father and mother still lived near Massena. When I told my boss of the opportunity on the Seaway, he wished me luck and said that my job with the Department of Defense would always be there if I wanted to come back. I remember he regretted that the Department of Defense didn't have the kind of money to allow him to pay me the same salary as the Seaway contractor was offering. Muriel and I put our house on the market and I accepted a position with Morrison-Knudsen. After they encountered some dredging issues, they closed that operation on the Seaway down for a year and transferred me to Salt Lake City, Utah, to work on the night shift on a rock-filled dam. After a week I told them I was going to quit. My boss, however, called the head office and they transferred me back to Massena to finish the dredging job.

Many of the operators, especially on the big dragline, the "gentleman," were older and very experienced. The guy who worked on it was about eighty years old. Badgett Mining brought that machine up to Massena in pieces and it needed continual maintenance. He was the only one who could take it apart, fix it, and put it back together again. He was quite a guy. Most of the other workers were transfers from jobs in the West. Others were long-time employees of various contractors. Merritt-Chapman and Scott had guys relocate from Connecticut who were dozer operators. They were nomads who went from one job. I had two assistants who had just finished their sophomore year at Clarkson University and had some surveying experience. One guy didn't go back the following the semester because he figured he was making good money and he could finish his degree when the project was over. Morrison-Knudsen hired him for a year and a half.

Jim Cotter

Contractors like Morrison-Knudsen and Peter Kiewit had workers who had been with them for years. So on the Seaway, they had a pool of skilled workers to bring along. However, B. Perini and Sons from Framingham, Massachusetts, who headed the construction effort on the Barnhart Island Power Project had never completed a hydro project before, so they didn't have that reservoir of knowledgeable workers. I don't know what other big projects they had constructed before that that even matched the size of the Seaway.

It took a tough supervisor to manage all of the different personalities on the job. On the Long Sault Dam, there was a supervisor with one bad eye that I had known from the Hungry Horse project. We called him "Walleye." He had a pair of binoculars and he would get up on some high point and survey the work and pick out the guys he wanted to fire. Most of my co-workers came from Colorado, California, and all over the West and many had worked on hydroelectric projects before, so their experience was what was important to Uhl, Hall, and Rich. There were a lot of TVA people who were recruited because of their surveying and dredging experience.

Alfred Mellett

I am originally from Wakefield, Massachusetts. In 1930 I passed the Massachusetts state highway supervisor's exam and stayed with them until World War II when I served in the Seabees in the South Pacific. I then became a civil engineer after four years of study at Massachusetts Institute of Technology and Lowell Institute. In 1953 I went to Turkey to work on a hydrodam project for a few years. When I came back to the United States, the St. Lawrence Power Project was starting. That was the biggest project I ever worked on. I initially served as a surveyor for Uhl, Hall, and Rich and then as the official photographer for PASNY. I had taken up photography as a hobby in high school and eventually turned it into a career with *National Geographic*. I was forty-three when I started working on the project and fifty when I left. Many of the supervisors were about my age.

The American section of the St. Lawrence Power Project extended for about thirty miles up and down the river. In 1954 I laid out and surveyed all of the big dams and channels with a bunch of surveyors. The next year construction started. Robert Moses, the chairman of PASNY, who was responsible for administering the power job, was also a friend of mine. He wanted a photographic record of all the phases of the construction sites. I told Robert Moses after the surveying was completed that I didn't have much to do. I informed him that I could take photographs and write weekly reports on all the project sites. He told me I had the job. I was given the use of an airplane and the funding to build a photo lab and hire three assistants to process the film. My pictures were sent to newspapers and dignitaries all over the world.

David Flewelling

I was born in California. During World War II, I lived near Syracuse, New York. In 1948 my family moved to New Hampshire, where I finished grade school and high school and entered college. In 1957 I was nineteen and I flunked out of college. As part of a northeastern cooperative program I had spent previous summers making a dollar an hour as a labor foreman and chief laborer. In March 1957 I hitchhiked from Pine River Junction to Massena, New York. The St. Lawrence Seaway was a big draw in those days because workers were in short supply and contractors were paying men high wages. I was hired the first week I was there by Perini as a draftsman on the Grasse River Lock doing rebar lift drawing. I was there from March until mid-July. When most of the drawings were completed, they cut back on the drafting staff. Since I was the last one hired, I was the first one fired. I went back the next week to collect my paycheck and the payroll office said the site manager wanted me to work on the Barnhart Island project, where I labored for the rest of the summer. In the fall I went back to college. Even though my time on the Seaway was relatively short, for a nineteen-year-old kid it was an eye-opening experience.

James Romano

I am originally from Rochester, New York. When the Seaway project started I was twenty-two and supplying gravel to the washing plants for the New York State Thruway. The Seaway contractors put ads in the newspaper encouraging all skilled and unskilled workers in the state to relocate to Massena and the surrounding towns. Other men working with me on the Thruway had told me of the great demand for truckers to transport gravel from the pits to the Seaway sites. I finally decided to move to Massena for three months in the wintertime. I was going to buy a trailer and live in it, but I didn't want my family to live like that. It was a bum's life. A lot of southerners brought their wives and kids on the road and followed the big projects across the country. People called them trailer trash. I didn't want my wife and daughter to be exposed to that lifestyle.

As the leader of A. J. Trucking, I approached contractors to see how many millions of yards of dirt they had to move. I had thirty trucks under my command. I didn't own them all, but I had thirty followers. In my own fleet, I had ten trucks and I hired the others. The contractors didn't want thirty guys to do the bookkeeping; they wanted to do business with one guy. To keep all of my drivers happy, I rotated the hauling. One day a few of us worked and the next day we parked. With this arrangement, all my drivers got equal time and equal pay. United you stand, divided you fall. If you worked on the job you made money. But you had to rent a room in a private home or boardinghouse or a hotel room and you had to eat. Trucking is a poor man's business.

I was hauling crushed stone from the off-site gravel pit into the batch plant. Each project site had it own concrete plant. The pit operators loaded three five-yard bins into the back of my truck. I would drive to the site I was assigned to and the crane operators would lift them out and place them outside the batch plant. For each batch of concrete the mixers would measure out the exact amount of gravel, sand, and liquid. The batch plant operators were instructed to only make a certain amount at a time. Then the laborers would dump the mix into hoppers that would be brought into the dam site by truck. The crane operators would then pick them up and dump them in the vibrating machines. The truckers would then bring back up the empty bins to be refilled.

There were license plates from all over the United States. Some of the guys talked about working on a hydroelectric plant near Las Vegas. Most of the southerners had worked on projects in the Tennessee Valley. They liked to drink a lot, but they kept to themselves because they thought we were just dumb Yankees. There were also a lot of truckers from nearby towns including Watertown. Local farmers and their sons got jobs as operating engineers because they knew how to drive machines. Other locals became carpenters because all they had to do to be hired by the union was be able to carry a hammer and a saw. Some of the guys also thought that they could just get in a truck and learn how to drive it. That is why in those days a lot of guys got killed.

The provincial government in Ontario selected Ontario Hydro as the lead agency on the Canadian side of the power dam. Besides guaranteeing the adherence to the project's time schedule and budget, Ontario Hydro had to oversee the work of their subcontractors. In the past Ontario Hydro had relied on their experienced pool of workers to complete their various hydro projects. But in the case of the Robert H. Saunders Power Dam, the only Ontario Hydro employees on-site were high-level administrators and supervisors. Iroquois Constructors assumed the day-to-day responsibility of overseeing the construction by numerous subcontractors and skilled and unskilled workers. Hydro personnel officers, however, retained their control over the recruitment of workers for the project for both themselves and their contractors and over overall inspection of concrete integrity and workmanship.

Ontario Hydro and their subcontractors completed their side of the dam by the 1958 deadline because of their experienced administrators and supervisors. On-site, Gordon Mitchell, the project director, dealt with the administrative and financial issues. Mitchell had earned a military cross for gallantry in World War II as a member of the Royal Engineers. He worked as a foreman on the waterworks construction in New York City and attended Toronto University. While he had supervised work at prior Ontario Hydro projects including Niagara Falls, he saw the Robert H. Saunders Power Dam as bigger in terms of cost and more complicated based on the number of houses and their occupants being relocated. Mitchell's right-hand man, William Hogg, also a veteran of the Niagara Falls project, advised Mitchell to surround himself with the experienced Hydro men and the best tradesmen from across Canada.[8]

Much like American contractors, Iroquois Constructors, Ontario Hydro, and their subcontractors employed a variety of skilled and unskilled workers.

Most of the skilled workers moved to the area from the east coast of Canada as did the equipment operators, supervisors, and inspectors. A large contingent of seasoned transient tradesmen who had worked on numerous projects for Ontario Hydro continued their employment for another four years. At the same time the hydrodam project began in Cornwall, Hydro finished up their Niagara Falls project and their administrative staff relocated to the area to work on the power dam. However, many Cornwall residents found employment with Canadian contractors as concrete workers, carpenters, and office personnel. They were trained on the job and later used these skills on local and national construction and public works projects.

Ontario Hydro workers also approached the residents of towns along the shore of the International Rapids, which were flooded to provide sufficient depth as well as the power pool, about either selling or moving their homes. This area included Iroquois, Morrisburg, Ingleside, and Long Sault, Ontario. The Canadian national government, along with the Province of Ontario, financed the relocation of 6,500 people to new towns built by Ontario Hydro employees. At the peak of construction in August 1956, Canadian contractors employed 4,550 tradesmen at the power dam, including 600 carpenters who erected forms, took them down, and rebuilt them for future use.[9]

George Haineault

I was a twenty-one-year-old kid from Cornwall when I was hired by Iroquois Constructors to work at the power dam site. I went to the field office and spoke with the personnel manager. He said that there was work for anyone who wanted a job. So I was hired. At the time, I had just completed my second year of an apprenticeship in carpentry with a local contractor, making $1.10 an hour. When I got hired by Iroquois I started at $2.50 an hour with time and a half on Saturday, and $5.00 on Sunday. It was big money for the time. Quite often I worked seven days a week. Being young, married, and the father of one child, I could use the extra income and I didn't have the ability to say no. Even though I spent my money wisely, when the project ended Hydro left and all of a sudden, after three years, I was laid off again; with no job, making ends meet was harder. It was the best job I ever had. I was amazed by all of the equipment.

Arnold Shane

I moved to Cornwall from Pendleton, Ontario, when I was twelve. I finished high school and then went to work full time on the Hydro job. In the spring of my senior year of high school, I heard that Iroquois Constructors was looking for truck drivers. Since I grew up on a farm, I had worked and driven tractors since I was ten years old. The Ontario Hydro personnel office liked farmers and farm boys because we knew how to work and how to operate the equipment. My friend who already had a job on the power dam site gave his trucking supervisor my name and that is how I got hired. I drove an eight-ton diesel dump truck during the excavation of the power dam site. I would pick up a load at the bottom of the eighty-five-foot excavation pit, drive up the dirt road to go and empty it, and go back down into the pit for another. The roads were steep and I didn't have a lot of passing room. The truckers hauled dirt all day and all night. I worked from six in the evening to midnight. I went to school all day, rode my bicycle home, ate supper, and biked to the power dam site, and drove a truck until midnight. I did that for four months until school ended. Even though I was only sixteen, I could drive a truck on-site as long as I did not go on the main road.

After I graduated from high school, I applied for an opening with Ontario Hydro and obtained a job in the stationary store. One day Neil Mustard, a concrete engineer from Toronto, came in to get his mail. He said he was looking for someone to assist in the concrete control lab recording all of the concrete breakdowns and doing other clerical work. He said he had checked my personnel record and I could start there the next morning as a junior concrete inspector. Back in those days, Ontario Hydro was highly regarded all over the world for their concrete knowledge and research. Contractors and governments in other countries called upon their engineers and designers to design concrete for their dams at the University Avenue location in Toronto. For six years after that I spent three months in the research lab, three months in the concrete batch plant, three months pouring concrete, and three months in school. It was the best education I ever had. By the end of the project, I was a concrete inspector and eventually became a civil engineer.

Donald Rankin ran the show for Iroquois Constructors. Don was probably in his forties when he started on the power dam project and had worked on hydro projects all over the country. Other workers came from Toronto, and the provinces of Ontario and Quebec. Some had worked on jobs like this before and some hadn't. Some of them stayed for the whole project, while others stayed for a few months and left.

Donald Rankin

I emigrated to Canada with my parents at the age of one from England. Prior to working on the power dam project, I had held many positions. I had been a miner in the gold fields of British Columbia and a topographer. As a member of a survey crew while at the University of British Columbia, I went into unmapped parts of the province searching for minerals. After I graduated from university, Dominion Hydrographic Service employed me to plot shoals and inlets of the British Columbia shoreline. I began working for Ontario Hydro in 1947 at the Pine Portage Development north of Lake Superior. In 1950, I became the division engineer in charge of the hydraulic tunnels at the Sir Adam Beck Niagara Generating Plant Number Two on the Niagara Falls project. Four years later when the St. Lawrence Seaway and Power Project began, a great many Ontario Hydro workers and staff, including myself, moved from the Niagara project to Cornwall. When I arrived in Cornwall in 1954, I assumed the role of divisional engineer at the power house project for Iroquois Constructors, which made me second in command to the project engineer. I was forty-five.

In my role as the divisional engineer for Iroquois Constructors at the power house project, I was involved with the construction beginning in 1954 when the contractors were surveying and preparing their offices for their employees. I became involved in the construction of the cofferdams, which dammed up the river at the power dam site. In my capacity as division engineer, the crews I oversaw dealt with the day-to-day matters on the power house construction. Every month Ontario Hydro accountants based contractors' compensations on their progress, including how much concrete was poured. I had to document the accomplishments of all of the subcontractors in my monthly reports for the project office, so each contractor was duly paid. I wouldn't say we were in charge of the project; we were the on-site coordinators. On the Canadian side there was a project manager and a project engineer who dealt with the financial and administrative matters. Eventually George Rainer, who had done a number of jobs for Ontario Hydro, took over supervision of the dam construction for Iroquois Constructors.

My crew members were all experienced and had worked on three or four other major construction projects. I had two assistants who were particularly important to me. The first was Jacques Linely, who was from overseas. This was his first job. The second fellow, Magnus Johnson, had come from Niagara Falls. I wouldn't say either one of those people had experience on power houses. I had a little experience, having worked on one power house previously. Most of the supporting group, instrument men and engineers, had worked for Ontario Hydro for a number of years. Immediately preceding the Niagara project, Ontario Hydro had been engaged in construction on the Ontario River. This resulted in my opinion in the construction division of Ontario Hydro having plenty of experienced construction workers to draw upon.

John Moss

I lived in Cornwall and I knew a couple of lads who were working for Hydro at the beginning of the power dam project. They told me that Hydro was going to be hiring some new employees. I went down to the employment office, filled out an application, and I got hired as a concrete inspector. I was twenty-six years old. I supervised the concrete pours that were assigned to me regarding their placement, moisture content, and slope. Ontario Hydro outlined certain specifications that each pour had to satisfy. I had not done this kind of work before. I had previously worked at Domtar, the local paper mill. An experienced concrete inspector trained me and told me what to look for. He had worked for Ontario Hydro in this capacity on a previous dam project. I worked with guys that had been with Hydro for several years and transferred to Cornwall from other projects. They were from all over Canada, including many from out West.

Keith Henry

I grew up in Alberta and went to the University of Alberta for my engineering degree. When I graduated from university, I had four written job offers to choose from. Ontario Hydro, however, offered me the best salary.

I moved to Toronto for eight years where I worked on the Ottawa River Project and in the hydraulic laboratory doing tests for the Niagara Falls project. I served as the river control engineer on the St. Lawrence project for five years. My appointment had to be agreed upon by the leadership of the International Joint Commission, the two Seaway authorities, PASNY, and the Corps of Engineers.

When the construction began, I spent two years splitting my time between Toronto and Cornwall. Once the dredging work was completed and the construction was well under way, Ontario Hydro officials decided that they needed a man stationed in Cornwall to keep track of how the various sites were affecting the river flow. Jack Bryce was in charge of finding someone to fill the river control engineer position as he had supervised the building of the models. They offered the job to my boss, Don Harkness, as he was the most qualified, but his wife did not want to move to Cornwall. Jack called me into his office in the fall of 1955 and asked if I would be willing to take the job and move to Cornwall. I jumped at the offer. I saw a career opportunity of a lifetime.

I worked late that night and didn't have time to mention the fact that I had accepted a new position in Cornwall to my wife, Marg. The next night we attended the annual Generation Department party. We had our first dance and then Harkness grabbed Marg and whirled her across the dance floor. He asked her how she felt about going to Cornwall. People were congratulating me and saying how wonderful she was to follow her husband to the wilds of Eastern Ontario. Marg agreed that it was a great opportunity for me, but has never let me forget that she found out of our impending move from Don Harkness. She was pleased to move to Cornwall because I had been spending more time there recently than in Toronto.

Prior to the Cornwall project, Ontario Hydro employees completed all aspects of a construction project from testing to the construction and inspection. When Ontario Hydro continually moved from one hydrodam to another, they always had new jobs to keep these guys busy. But when they finished the job, they basically couldn't fire the people that were on their construction crews. Dick Hearn, who became the chairman of Ontario Hydro after the death of Robert Saunders, had begun his career with Ontario Hydro, and then went to work with different outfits out

West as a consulting engineering. He realized that one entity could do the design and engineering work and then effectively hand off the actual construction to outside contractors. He instituted this practice at Ontario Hydro during the Cornwall project. Ontario Hydro held onto the supervisory work and left the construction work to outside contractors.

Many of the skilled workers labored on the power dam for a few months and then went back home. Others stayed for a number of years because they had such good jobs and were earning good wages. For about a five-year period, the construction went full blast. The local people got more jobs as the time went on. They turned out to be good workers and, of course, since they already lived there they didn't want to leave.

Arthur Murphy

I was born in Cornwall, Ontario, but raised in Montreal. My dad grew up on a dairy farm in Dickinson's Landing, one of the flooded communities. After eight years, my family moved back to Cornwall where I finished grade school and entered high school. During my summer vacation from high school, I worked part time in Ontario Hydro's construction warehouse logging in sections of cranes and machinery. When I graduated from high school, the directors of Ontario Hydro's information section, now known as the public relations department, hired me as a tour guide to tell the story of the power project to the thousands of onlookers who visited the power dam site daily. As my personality was conducive to that kind of environment, I eventually became the chief guide at the heart of the project. I oversaw a staff of university women who served as tour guides. They were hand-picked, not only for their knowledge and communication ability, but also for their good looks. Each lady needed to make a really good impression on our visitors.

Our office was in what is referred to today as the penthouse at the generating station. We showed a film depicting the construction activity in our big auditorium and then took people on tours. Some days we recorded over four thousand visitors. In 1957 Hydro completed a new reception center which housed displays of project models, pictures of the construction, and historical information on the power dam as well as serving as a starting point for all my guided tours.

Bart Whitten

I was a geological supervisor for four years and two months on the power dam project for Ontario Hydro. I am originally from Renfrow, Ontario. The Ontario Hydro personnel office hired me when I was eighteen to work as a carpenter on the Niagara Falls project. Six years later, when I was twenty-four, I was transferred to Cornwall to work on the power dam project. Ontario Hydro brought me in to check on their contractors' progress and ensure they followed Hydro standards and plans in terms of concrete pouring and earth removal. I certified their work at the end of the shift and documented everything that happened. My first assignment was on the sixty-cell steel cofferdam which was exclusively built by Canada.

In terms of the other men on the project, while many of us had been employed on other Hydro projects, others had no experience. The locals got word that contractors often needed workers at the spur of the moment, so men would camp out at the project's main gate near the main highway. If the contractors needed men, they went to the gate and got them. The shift supervisors paid them cash, so as to not alert the personnel office to this temporary workforce. Eventually some of these men obtained full-time jobs. Most of the Hydro chaps worked as supervisors in their own specialties and had traveled all over Canada to oversee various dam and waterway projects.

Joseph Couture

I came to Cornwall in July 1954 as an employee of Ontario Hydro. My boss transferred me to work for Iroquois Constructors, the main contractor on the power dam project, as a carpenter. I enjoyed working on the project immensely. I started as a carpenter and worked my way up to foreman. I was the youngest carpenter and foreman at twenty-four and twenty-six respectively. Initially, as a carpenter, I built the forms for the concrete pours, and then as a foreman I oversaw the upstream portion of the dam, also known as the whole Canadian side of the dam. I had been employed on projects like this before for Ontario Hydro at Niagara Falls and in Winnipeg. I found the work to be very satisfying and I had great respect from my supervisors and from my workers, except the ones I fired.

I was born in Quebec and had been working at an aluminum plant in British Columbia for four years when I got the call to work construction. As a young man, I liked the big jobs because of the overtime. I learned my trade on the job from some of the best carpenters. A limited number of men were trained to do this kind of work. Three or four foremen did not speak English and many of the carpenters did not speak French. At the beginning, they could not understand each other, but by the end of the project they had learned to talk about the job in each other's language. I spoke both French and English, which gave me a great advantage because I was one of the few foremen who did.

Les Cruickshank

I was born in Quebec in the Ottawa Valley west of Ottawa where my Scotch and Irish ancestors settled. They had come over during the potato famine. My great-grandfather purchased a piece of land with his two brothers. He kept the middle tract, cleared it, and built a big family house. My grandmother, my mother, and I were born there. We lived right along the Ottawa River, which is in northern Quebec and undeveloped. The normal course for young fellows back then was to head for Ottawa on the Ontario side to seek employment and that is what I did. Ontario Hydro engineers were building a dam that was thirty miles away from where I lived. I went over and got a job as an equipment operator.

I worked for Ontario Hydro from 1946 to 1956 on dam sites in various parts of the province. Prior to relocating to Cornwall, I was employed by them at Niagara Falls where they built a large power house by taking water from above the falls and converting it through a couple of tunnels and an open canal below the falls. When that project was over, Ontario Hydro administrators decided that building their own dams with their own construction crews was no longer cost-efficient, so they decided to hire outside contractors for the construction phase of future projects beginning with the power dam in Cornwall. So I traveled from Niagara Falls to Cornwall in December 1955 and talked to the rehabilitation project engineer, whom I knew from other projects. I asked him if he would hire me if I bought a grader. He said he had to wait and see what his needs were in the spring. I went back a second time in March 1956 and he agreed to hire me for the next two years. So I bought a grader and worked at the village of Ingleside for Ontario Hydro, which was township number one, for a year and a half. I earned enough startup money for my own business. In January 1957 I got married and purchased a house in the area. I bought several more machines, and my dad, who was an old builder and had worked throughout Ontario building roads for various contractors, came down and taught me how to build roads.

When Ontario Hydro came into the area the only workforce available was farmers, a few skilled tradesmen who had worked on local construction projects, and assembly line workers from the three area manufacturing plants. The unions set up apprenticeship programs to train carpenters, electricians, plumbers, and equipment operators. In the construction business there were a lot of floaters. They would leave and go home and try another job and then come back to the Ontario Hydro project. As the saying goes, the grass is greener on the other side of the fence. When one hydro project ended, they would follow Ontario Hydro to the next site. If they heard there was a big construction job going on somewhere they would go and see if they could get hired. The foremen and supervisors on the project were generally a mix of long-time Hydro employees and new junior engineers. Local people made up the rest of the skilled workforce and they eventually become good tradesmen. That is how they got their initial experience and apprenticeships. I don't think there were any local people that wanted a job that didn't get one. The project offered many local men the opportunity to learn a trade.

William Rutley

I was born about five miles north of Ingleside in the sixth district of Osterbrook. At the age of twenty-two I wanted to do something else besides farming and the Seaway construction began. I put in an application at the unemployment commission and Ontario Hydro hired me to work on the town sites. As an engineer's assistant for Ontario Hydro, I measured houses in Iroquois, Ontario, and determined whether they should be moved to one of the two townships. I ran parallels on the front, back, and sides of the house to see that it was all square. If it was in too bad a shape, they didn't move it. When I wasn't working I was at home helping my brothers on my parents' farm. In the winter of 1957, my shift started at 6:00 A.M., so I was up at four looking after fifty head of cattle before I went to work.

Because of the depressed economy in Nova Scotia and Quebec at the time, a lot of workers came from the east coast of Canada. Cornwall residents and those from surrounding counties could fill contractors' needs for carpenters and electricians, but for something this big, they had to recruit workers from all over Canada. Locally, there weren't ten thousand skilled and unskilled workers available.

A few refugees emigrated to Cornwall after World War II, which I referred to as displaced persons. These men were skilled tradesmen in their native countries who had bought land and become farmers when they came to Canada. Many were having a rough time financially, so they worked on the power dam for extra cash. These men had young wives and young kids and farming was a lot of hard work, but they eventually succeeded.

Glenn Dafoe

I grew up on a farm four miles outside of Aultsville, Ontario. When I finished high school, I decided that I wasn't an academic and wanted to work with my hands, so I elected to learn a trade. I became a carpenter's apprentice for four years before I was married. I got a job as a carpenter for Ontario Hydro on the Seaway project through the union. I started on the Iroquois Lock and then moved to the town sites. I put in the basement for the first house that was moved. I spent the bulk of my time in between Iroquois and Morrisburg involved with the reconstruction of the town sites, building the concrete forms for the sidewalks, street curbs, and foundations for the new shopping plazas. I was twenty-four when I started working on the project in 1955. I ended up working there for three years. The Seaway was an employment opportunity that hadn't presented itself earlier and never did again.

I worked with a few men who came in from outside, but the majority of the carpenters, electricians, masons, and laborers on the town site rehabilitation projects lived within a forty-mile radius of the job site. At the outset, many workers from the Niagara Falls project were given preference in terms of hiring, but many of them left after a few months because they were tired of being separated from their families. There was such a demand for manpower that some men learned new skills or bought equipment specifically to secure a position with Ontario Hydro or one of the numerous subcontractors. For example, there were a number of locals who had no trucking experience. However, they bought dump trucks and got hired by one of the contractors to haul gravel from one of the off-site pits to the town sites. I also knew a man who came from Nova Scotia and operated his own small backhoe. Many of the engineers and inspectors were experienced men in their fifties who had completed many hydro and navigational facilities for Ontario Hydro.

Hubert Miron

I worked for Ontario Hydro from 1956 to 1959 as a payroll clerk. I was thirty years old when I was hired by Bill Goodrich, the personnel officer, after I had filled out an application. I was born in Cornwall and had no previous accounting experience. My first job was working at Courtaulds. I also owned a smoke shop and billiards hall for fifteen years.

Barbara Brennan Taylor

I grew up in Cornwall, and after I finished high school I worked for three years as a secretary at Courtaulds, the local textile factory. It was a very stodgy British company and I felt I had no room for advancement. So I went down to the Employment Service Office and applied for a job on the power dam project. They sent me out for an interview with Iroquois Constructors, the main contractor on the project. I was twenty-two when I started working as the office manager's secretary. The working conditions and people at Courtaulds were so different than what I encountered on the Seaway. It was like going from one extreme to another. The workers and my boss at the textile plant were very boring, while on the Seaway we just had a lot of fun. I was there for three years until 1958 when I got married.

I was the secretary for the office manager, but I filled in on other positions if someone was sick. I was responsible for the normal secretarial duties, but also, because there were only ten of us, if somebody was missing at the switchboard or in another part of the office one of us filled in.

I remember being on the switchboard one day and this German engineer called and I couldn't understand what he was saying. I was not Bell-trained as our regular operator was and I kept asking him to repeat himself and he got very annoyed. If I had been working for Courtaulds and said what I said, I probably would have been fired. I told him if he took the marbles out of his mouth I might be able to understand him! That is a terrible thing to say to anybody, but he kept acting like I was really stupid because I couldn't understand him. I also wrote a gossip column for the local paper. There weren't any great scandals. I just know I had a great time writing it. My friend at Hydro wrote a similar column. It was just a fun time. I used to get calls when I was writing that column from men who didn't want me to put their story in the paper. They would say, "My mother gets this paper in Ganonaque and I don't want her to read about that." If anybody had a car accident, we wrote about it. I would just write up who was seen at what party with whom. It was a gossip column more than anything else. Everybody loved it.

I remember what we in the office thought was a funny story. One of the girls in payroll got called by a construction supervisor. He told her he needed twenty-five men for the graveyard shift, so she called the unemployment insurance commission office and asked for twenty-five gravediggers. She never lived that down. But they were moving graveyards along the Seaway project. She had never worked in this field before and did not understand what the man wanted. We ribbed her about it forever.

The building I worked in was a temporary aluminum shack west of the city. At one point they had a driver named Rex who picked all of us girls up at our homes in the winter and drove us to work. It was cold in the winter and hot in the summer and there was no air conditioning. I could see through the cracks in the floor. I used to wear a skirt or a dress every day. I worked from 9:00 A.M. to 5:00 P.M. I don't think I ever took a lunch except in the summer when a few of us went swimming. There was no cafeteria on our job site, so I brought my lunch. I was also paid very well because of the conditions.

A lot of the workers came from Quebec to work on the project. Some of them came into my office and talked about their wives and their children who weren't there. I suppose this was because many of them lived with men on the project, so any woman that was open to listening to their stories was really appreciated. Basically they talked about their home life and how lonely they were. Some of them came in to talk, while there were others who were on the make. They often talked about other construction sites they had worked on.

The St. Lawrence Seaway Authority was established by an Act of Parliament in 1954. The administrators and employees of the new agency were charged with completing all of the navigational work on the Canadian side of the Seaway. These responsibilities required the acquisition of land to be flooded, the dredging of the entire waterway between Montreal and Lake Erie to a depth of twenty-seven feet, along with the building of new locks and dams including the Iroquois Lock to bypass the American Iroquois Dam. The massive effort included the removal of 192.5 million cubic yards of earth, the pouring of 5.7 million cubic yards of concrete, the building of fifteen miles of dikes, and the digging of sixty-eight miles of channels.[10]

Garry Moore

I am a native of Cornwall, Ontario. In the summer of 1955 I went to the St. Lawrence Seaway Marine Division office at the end of Augusta Street and told the office manager that I could drive a motorboat and that I was an excellent swimmer. I also told him that I didn't have any work experience, but I asked him for a job anyway. I was hired on the spot. My first job was transporting inspectors back and forth to the dredges. That was a great job compared to some of my contemporaries who were working at the paper mill (Howard Smith). In June 1956 I graduated from high school and I worked on the Seaway project for another year before I quit and went to university. I had finished grade thirteen which meant I had learned trigonometry and geometry and was able to apply that to surveying. I worked on a hydrographic survey crew with the St. Lawrence Seaway Authority sounding and putting up ranges for the dredge line operators. My final assignment during construction was surveying the dredging of the Seaway channel in Lake St. Francis, down through Beauharnois and into Lake St. Louis. In 1961 I took a permanent position with the Canadian Seaway Authority and worked for thirty-five years in Cornwall, St. Catharines, and at the head office in Ottawa.

Most of my surveying training I obtained on the job. In the winter of 1956 the U.S. Army Corps of Engineers set up a university-level surveying course offered by Professor Hudson from Clarkson University at the Waddington High School. The amount of tuition that the Seaway Authority and the Corps paid varied based on the number of students registered. The more workers who took the course, the cheaper the price. So they asked the Canadian Seaway officials if anybody on our side of the border wanted to take the course and four of us went. We drove over the border one night a week and listened to lectures. I benefited so much from that course because I went to work the next morning and applied everything I had learned the night before.

One of the most important institutions to a worker in terms of hiring and maintenance of employment was his union. The location of the American construction sites in economically depressed areas with no previous large building projects explained the absence of strong building trades organizations. Thus Laborers' Local 322, an affiliate of Carpenters' Local 747, and Painters' Local 37 established hired halls in Massena at the commencement of the Seaway and power dam construction. Before being assigned to a contractor, workers filled out applications at these facilities and paid their annual dues. Each day unemployed workers arrived to be assigned to a construction site or a contractor based on the daily job list compiled by the hiring staff. These organizations also negotiated generous wages for their members because of the shortage of skilled labor in Massena. All contracts included clauses that obligated contractors to adhere to the established pay scale for each trade, which averaged 50 percent higher on the U.S. side, and made labor union leaders promise to not hire workers from the other side of the border.

The leaders of the Laborers' International Union of North America established Local 322 in Massena in July 1954 to meet the various contractors' needs for union laborers on the Seaway and Power Project sites. Previously, the union's Syracuse, New York, office was the closest local. If a particular contractor required the services of laborers in the area before that time, he would directly hire either nonunion workers or relocated members of the Syracuse affiliate. At the peak of the Seaway and power dam construction, the local's membership consisted of over six thousand laborers.

The chapter's founder, Sam Agati, a native of Syracuse, was described during the time of the Seaway project as "in his early 30's, citified, but dressed in a dark suit, shirt and tie, like an up to date insurance salesman."[11] Business agent Robert Ashley played an integral role on Agati's staff. He met with contractors and determined how many workers were needed on a particular site and for how long. Then he oversaw the screening and acceptance of these workers at hiring halls and the assignment of these members to particular sites. Ashley also smoothed over differences between workers and supervisors and attempted to make the atmosphere on and off the construction site one of harmony rather than of hostility.

Sam Agati

The project was big enough to warrant a local union with its own autonomy. It would have been impossible to run the show out of the Syracuse office. Many members wondered why the national had not put a local man in charge. The basic reason was that he would not have had the expertise to get the office and hiring hall up and running and deal with the demands and dominating personalities of the contractors and their on-site supervisors. The fact that I wasn't from the area made my task of organizing the union a bit more challenging. I had a tough time being accepted by the membership because I was a city guy and most of the members were locals. That's where my stewards really helped me out. I hired local people as my stewards, and with their help and the large amount of Italian workers, I was able to win the support of the membership.

In hindsight, I probably could have tried to blend into the community a bit better, but that went against my outgoing personality back then. None of my actions, no matter how small, went unnoticed by the local newspaper. Massena was such a small town and everyone knew each other. Whenever I walked into a restaurant, everyone knew I was a stranger. I definitely stuck out because of the car I drove and the way I dressed.

In order to initially join the union, a man needed to pay the $10.00 down payment and fill out an application at the hiring hall. Once the hiring agents found a new member employment, $10.00 a week was deducted from his paycheck by the contractor's payroll department and forwarded to us until the $60.00 membership fee was paid. After that all members paid union dues of $3.00 a month.

All of the laborers who worked on the project had to be a member of the union. Contractors could not openly advertise laborer positions in the newspaper as they had signed a contract with us that the project would be run as a closed shop. They all had to go through the various trades union hiring halls back in the fifties. They couldn't hire your average man off the street. Some contractors hired our members for four years, while others requested men for only a day. No matter, each of our laborers had to be paid the negotiated union wage.

I had a lot of clout as far as solving problems on the various job sites. If I thought my members were being treated unfairly by their supervisor or foreman, I had the ability to shut down the construction at that particular site. If one of my stewards uncovered unsafe working conditions or other issues, I resolved the situation by either speaking directly to the managers of the contractors' labor relations departments or calling the company's owner versus dealing with their site supervisors. For all union contractors, dealing with unions was a way of life. The employers were frightened of us because they knew that the union might hold back qualified workers from the project if they refused to cooperate with us.

Jack Bryant

The unions had quite a tight control over the contractors and all of the agencies involved. You had to buy your way into the local and that might not have been popular with some people. Even though all of the unions had hiring halls, some workers always suspected that the union stewards did not make the jobs available to all members equally. They played favorites, but that is not unusual. That used to happen on every job. I also think that [there] could have been more strikes on the project, but a lot of the time contractors and union officials settled their differences with a wink and a nod from Power Authority officials, saying they would compensate the unions later because they had to get the job done. All of the contracts supposedly included a no-strike clause.

Roderick Nicklaw

David Manley and I were both from Malone. We had some interesting experiences together. I lied to get a job on the project. I had to. I was supposed to be eighteen, but I was fifteen when I first went to work there. I didn't look old. I looked young, but I could do a man's day's work. David and I went to the laborers' union hall and paid our dues. If you went and paid dues, they didn't ask how old you were. They just sent us to work. I went to school during the day, played football after school, and worked on the Seaway at night. I didn't get much sleep. I slept in the car going back and forth to work. I also slept a lot in class. I had always worked on a farm as did Dave Manley and the rest of the guys that worked there. Our parents didn't have any money to give us, so we went out and earned it. I worked on a crew of all young guys. You didn't have to have a whole lot of brains to be a laborer. If you could use a shovel and a rake and had a strong back, you could be a laborer.

Johnny Burt was the superintendent. He was a rebel. He was from down South. He was a nice guy. All of the bosses wore Masonic rings. If you weren't a Mason, you weren't a boss. He had worked for Morrison-Knudsen for many years. Many local people became regular supervisors, but not superintendents. Burt oversaw the three crews pouring concrete on my shift and the signal man guiding the cranes for the pours. They had a superintendent on each shift and they would have several foremen. We had so much to do, we didn't need a foreman. At the beginning of every shift, he told us what we had to complete during that time frame.

David Manley

I was a junior at Malone High School in 1956. Four of my friends and I knew the laborers were hiring because a lot of the southerners were not back yet. Many of them returned to their homes in the winter months. I went to the laborers' union hall in April, filled out some papers, and then went and stood outside. If a contractor needed a laborer that day, they called my name and told me which site manager to report to. This was a daily process until I was hired permanently. I also had to buy a union book for twenty dollars and sign a form so they could deduct my dues from my weekly paycheck. I was fifteen when I started. I lied about my age. The union leadership overlooked the fact that I was underage because the hiring hall could not supply the contractors with enough workers. Many of the southern men did not like to work during the colder months or even during the summer on the midnight shift, so I got a job working midnights. Then when school was out, I worked four to midnight six days one week, and seven days the following week. I was young then.

I started off as a laborer and then volunteered for concrete work because that was ten cents more an hour. It was hard and dirty work, but being a young kid, I didn't mind it. I really enjoyed it. I worked on the power dam for a while, and then on the excavation of the Snell Lock. The contractors on that site blasted all the rock and removed all the clay to create the base of the lock. The following year I went back and worked on the tunnel of the Robinson Bay Lock, now known as the Eisenhower Lock.

There were a lot of men from Georgia, Alabama, Florida, and Virginia. It was a shock for them to come to Massena, but they did enjoy the fishing. It was cold at night for them even in the summertime. Some of the workers were locals. I worked with some Indians from the reservation who were steel workers and concrete vibrators. There were a lot of local farmers who were hired by the contractors to operate machinery. Alcoa

12. Concrete buckets are transported from the Eisenhower Lock batch plant to the waiting cranes to deliver to the concrete crews who are beginning to pour the foundation. Courtesy of Cornelius J. McKenna.

had also laid off a number of employees in 1953 and many of those men found jobs at the lock and dam sites. For them, the project couldn't have come at a better time.

Frank Wicks

I was born in Watertown, New York, and grew up in Canton, New York, where I finished high school in 1957. The laborers' union steward, William Tisdall, lived on State Street in Canton. Someone told me he was the guy to talk to if I wanted a job on the Seaway. I went and spoke with him and he directed me to go to the laborers' union hall in Massena, where I filled out an application. Soon after, I was hired to work as a member of a concrete crew for Uhl, Hall, and Rich with mostly Indians on the power dam in the summer of 1957. I started out as the hoser downer on the prep group and then progressed to a vibrator. I worked from four to midnight, the swing shift. Most of the guys I worked with carpooled, but I drove myself so I could stop at the visitor's center every day and look at the model of the entire project. When I first saw it, I didn't comprehend how it functioned. But I kept going back and studying the layout, so I would say I became an expert. All of the guys I worked with were older than me as I was eighteen and they were in their thirties and forties.

Ted Catanzarite

I graduated from high school in 1955 and worked on the Seaway for two years until I entered university in September 1957. In the winter of 1955 I worked at the power dam solidifying the bedrock. We drilled holes into the bedrock and filled them with liquid concrete, called grout. In the spring of 1956 I chased one of the supervisors around the project for a while and got fired. I was a young guy then.

Many of the contractors and their workers were from the West and the South. I am from an Italian background and the outsiders had a certain attitude toward people of Italian descent and I was sensitive to that. They used to call me a "goomba." I was a little bit of a hothead. After I was fired, I worked for a short time with a civil engineer at another site. When my old supervisor heard that I was still working on the project, I got fired again. Then I found a job at Alcoa for four months doing inventory.

At the same time I got in trouble with the laborers' union. I said something at a meeting and they kicked me out too. I ended up making peace with the union in the fall of 1956. One of my cousin's good friends, Robert Ashley, was a big shot in the union. I gave him a case of good scotch and had to promise not to go to the meetings any more. He, in turn, told the men at the hiring hall to put me back on the list. I then worked on the Iroquois Control Dam and the Eisenhower tunnel until August 1957 when I went to university. I initially worked on the midnight shift. I had to work my way back in slowly.

13. Ted Catanzarite in front of his parents' garage before leaving for work in August 1957. Courtesy of Ted Catanzarite.

Thomas Rink

When I graduated from high school in Hornell, New York, at the age of seventeen, I started working for Uhl, Hall, and Rich. My brother-in-law, Joe Marmo, was already working on the Seaway and helped me get a job. There was so much work, the contractors were just hiring anyone they would find. They were hungry for workers. My brother and I moved to Massena and lived in a house with my sister Ann and her husband Joe. Some carpenters and laborers were hired on the Seaway project and only worked for two weeks because they didn't like being away from home. I worked for three years from 1955 to 1958 as a laborer. I was very fortunate. I went from one job site to another. I had no experience doing this kind of work before. There were all kinds of people on the Seaway project and we all had to get along. We didn't form cliques. We were all equals working toward the same goal.

One of my first assignments on the project was testing soil and densities. I had a leak in this little shanty where I had my soil test equipment, so I told my supervisor that I wanted to get some rolled felt paper to fix the roof. During the four o'clock shift, I started nailing it down and some guy asked if I had a carpenter's book. When I said, "No!" he told me to get the hell off the roof. He then brought a carpenter crew in and put on a whole new roof for me.

The locals all ran a pretty stiff union job. You found new work assignments through your union, you paid your dues and got your paycheck. That was the way it was. I can remember going to the union hall, because they made me a union steward. There was a list of positions that needed to be filled on the different project sites. It always seemed like they were trying to fill the void because the contractors would request twenty or thirty men and they just weren't available.

During the Seaway and power dam construction, officials of Carpenter's Local 747, headquartered in Watertown, New York, set up an affiliate in Massena. As was done for the laborers' union, business agents established a hiring hall to process contractors' employment requests, accept workers' applications, and supply the needed number of carpenters to each site. While many carpenters had worked on private construction jobs including building schools and homes, most had never assembled forms to shape concrete pours on a large dam or lock.

Neil McKenna

I was born in Potsdam, New York. From a young age I was fascinated by anything that had to do with construction, trucks, earthmoving equipment, and concrete. Therefore, when the contractors started hiring for the Seaway project, it was not an opportunity I wanted to miss. I worked as a carpenter on the Long Sault Dam, the Grasse River Lock, now known as the Snell Lock, and the Eisenhower Lock between November 1956 and December 1957. The carpenters' union steward, Mac MacDonald, was my next-door neighbor in Potsdam. So he knew my father pretty well. Mac sent me to the carpenters' union hall in Massena to fill out an application and I was hired instantly because of his referral. I was working on a ninety-foot-high dam in a week's time. It was exciting. The large equipment and the massive concrete pours were all brand new to me as a recent high school graduate.

I began my year-long stint on the Grasse River Lock project as a first-year apprentice carpenter for Morrison-Knudsen. I had no more business having that position than the man in the moon. The U.S. Army Corps of Engineers eventually hired me as an instrument man and notekeeper on a survey crew. I went around and checked all of the survey markers placed by the contractors' engineers. I got a lot of good experience and learned a lot. It was very good pay right out of high school. Of course, it didn't last. I remember my father telling me, "You have a good job now, but soon you will be unemployed."

The surveyors I worked with had gained experience on other large dams and big construction projects. However, most of the carpenters were from Massena Center. The resident engineer for the lock for the Corps was a graduate of Clarkson University. Other engineers and inspectors transferred from the TVA and the Bureau of Reclamation and had worked on the Shasta Dam and the Snake River Dam. My party chief, Harvey, came from the Shasta Dam. Every time we started a new phase of the lock, he talked about how he had done the same thing on other projects. He indicated that although this was an entirely different kind of lock and dam, the skills needed and the basic design elements were similar. Harvey also

14. Cornelius (Neil) J. McKenna Sr., Warren Scofield, and Harvey Day discuss survey work for the day at the Eisenhower Lock site. Courtesy of Cornelius J. McKenna.

invented a surveying measuring stick to attach to the south wall of the Eisenhower Lock to align it with the north side. It was my responsibility to nail it onto the finished south wall along with a friend of mine. I must have been crazy!

Ira Miller

When the Seaway construction began I was driving a milk route and hauling pulp with my own truck. When I was hired on the Seaway as a carpenter, I gave up trucking, but kept the milk route. One of my friends told me the carpenters' union was hiring. So I went to the union hall and the business agent told me if I paid the $100 annual union dues, I could start working that night. My only carpentry experience was repairing my own home, barn, and garage. On the Seaway, I set up concrete forms and tore them down after the concrete had cured. My first assignment was at the very bottom of the Long Sault Dam and when that was done I went to the Barnhart Island Powerhouse. I had a 1952 Mercury that I drove from my family's homestead in Nicholville, New York, to the job site every day.

My co-workers were from every state in the Union. The older engineers and inspectors had started their careers on the Boulder Dam. Many of the local guys had done some type of carpentry before they were admitted into the union. There had been a lot of schools built in the area and that gave a lot of these guys a little taste of large construction. However, the Seaway was a different type of carpentry work. The material was a lot heavier and the pace was faster than on a regular construction site. There was definitely a strict timetable that we had to abide by. As long as all the forms were properly constructed and placed and the pours were level, your work passed inspection. Uhl, Hall, and Rich engineers were constantly checking the forms, the pours, and all the steel work on the Long Sault Dam and the main power dam. If they found a mistake or something that differed from the original plans, we had to fix it.

The painters and masons served important roles during the construction of the locks and the power dam. As a rule, before a worker was hired by one of the numerous painting and masonry contractors, he had to either join Local 37 or Local 81. These men faced many unique dangers on the job. As they covered every exposed metal surface on the various sites, they inhaled the toxic fumes given off by the paint and often worked at extreme heights and through the night to complete their tasks before the parts could be installed by other trades. Masons often suspended themselves in awkward spots high up on the lock walls to fill bolt holes and seams. Many also performed more aesthetic work on the administration building, whose facade was covered with sandstone that was difficult to work with.

Ron Barkley

I am from Waddington, New York, originally. When I came home from the marine corps, there was no work. The only jobs available were with the contractors completing the Seaway and power dam sites. I was hired by V. A. Roberts, the project manager for Morrison-Knudsen, at the Red Mills section of the Galop South Channel improvement. Almost one million yards of earth had been removed before I got there, using two large draglines and nine small shovels which scraped the dirt and dumped it into fifteen Euclids. Only one of the draglines remained when I was hired.

15. In March 1958 the Long Sault Dam is complete, and water from the raised power pool flows over the spillways at each end of the structure. A section of the suspended bridge that had been used to transport supplies and workers still remains to be dismantled. Courtesy of Alfred Mellett and PASNY.

The operator was pulling out the remainder of the dike and removing the last of the soil from the river bottom. The dike was meant to prevent the area known as Red Mills near the Iroquois Control Dam from flooding. I drove a truck and a Euclid and took the dirt off-site.

I was then hired by J. I. Hatt to paint the gates, the gantry cranes, and all the steel on the Long Sault Dam. For that position, I joined the painters' union. The dam was huge. I had some pretty scary moments. The winds would pick up quite regularly on the river and before I knew it, it was blowing at fifty miles an hour. It almost blew me off. No one fell off or got killed.

My various positions on the Seaway were my first civilian jobs off the farm. Before I went into the marine corps, I worked for five months at Alcoa. I had not been paid that kind of money before. It was amazing. I could buy anything I ever wanted. I was a country boy working with experienced painters and truckers who were in their forties. It made me realize that being a painter on large construction projects could be my lifelong career. When the Seaway was completed in 1959, I moved on to the Niagara Falls project.

Thomas Sherry

In 1956, when I was twenty-five years old, I was hired by Perini through the Masons Local 81 to work on the Eisenhower Lock. I had never worked on such a large construction project. The business agent charged me $5 to drive me out to the job site. At that time that was almost two hours of wages. So I asked him what the $5 was for and he said that it was not something official, it was just to pay for his beer after work. That kind of stuff was going on. He got mad that I questioned him. I spent two months as a mason on the Grasse River Lock working for Morrison-Knudsen. They were one of the biggest construction companies worldwide.

My brother-in-law was a farmer and he wasn't making a lot of money, so he quit and worked as a truck driver. There were a lot of unskilled people. The guys who called themselves carpenters built concrete forms and screwed in bolts. There was nothing technical about it. If they paid the union dues, the steward gave them a carpenter's book, a crowbar, and a hammer and made them carpenters. There were a lot of what we called "Seaway carpenters." Once the project was done, they couldn't get a job because they didn't know what they were doing.

Most of the superintendents from the West were all smart. They had worked on all these big projects and dams all over the country. It is a good thing that they were there because they taught me how to use the equipment. They taught novices, like me, how to work safely. Everyone else just did what they were told to do. They just needed bodies. I think the contractors were required to have a certain number of men on the payroll at a time to get paid by the Power Authority or the Seaway Development Corporation.

Members of Operating Engineers Union Local 545 performed the most challenging tasks on the project sites and operated expensive equipment. Since the union's main upstate New York office was located in Rochester, more than one hundred miles away, officials decided to create a new affiliate in Massena. They sent their union leader from Rochester to establish and administer the new facility until the completion of the Seaway project. Before a machine operator or oiler could join Local 545, he had to work on the project for a few weeks, so union officials could access his skill level. If his operating ability was deemed acceptable, he paid a $125 membership fee, in order to become a permanent member. For the duration of the project oilers and operating engineers paid $4.00 and $7.00 in monthly dues, respectively. Members of other locals and nonunion men were charged licensing fees of $2.50 per week or $10.00 a month.

A strike by members of the International Union of Operating Engineers on March 12, 1956, stalled the entire project. Picketers from Local 545 blocked the entrance to all of the work sites, and members of other unions did not attempt to cross the line. The afternoon before as rumors spread of the impending strike, contractors shut down and secured equipment at 4:00 P.M. to prevent sabotage. The operating engineers had been working without a contract since their last one expired in December 1955. The disagreement concerned the merits of a carryover clause in the contract that stated that for the life of the contract, the wages would remain at the same level. Al Miller, the local's business agent, called a meeting and informed his members that work would not resume until a contract was signed that included a thirty-cent pay raise over the next two years. The Associated General Contractors, the negotiating agent for the contractors, had made an offer of a ten-cent increase to those workers employed on the Eisenhower and Snell Locks because of the 1956 bid date. The group had managed to settle contracts with the other 1,600 workers and their unions and after a one-week strike offered the operating engineers a twenty-five-cent raise across the board effective January 1, 1957. The contract also included wage adjustments for operators of special types of heavy equipment.[13]

On December 11, 1959, the owners of Morrison-Knudsen, Walsh Construction Company, and B. Perini and Sons sought the review of a March 4, 1959, order of the National Labor Relations Board (NLRB) regarding the companies' unfair labor practices related to the use of Local 545 as their employment agency for operating engineers. None of these contractors established a human resource office in Massena and therefore relied predominantly on the Local 545 hiring hall to supply them with oilers and operating engineers. The Board found that since March 1, 1956, the petitioners had entered into an exclusive hiring arrangement with Local 545 and the International Union of Operating Engineers, which resulted in discrimination against nonunion workers. When the site supervisors at the Eisenhower Lock or Massena Intake needed more oilers or machine operators, they compiled a list including the exact number and skill levels required and relayed it to Local 545's hiring agents. These men in turn consulted their list of available workers and sent them with referral slips to the appropriate company's on-site representative. If a qualified man tried to avoid the union hiring apparatus and applied at the actual site for work, a foreman sent him to the hiring hall to obtain a referral slip or directed him to the operating engineers' on-site union steward.[14]

The NLRB members ruled that giving preference to members of Local 545 over members of other locals of the International was a violation of the National Labor Relations Act because encouragement of membership in a particular local was unlawful. The board ordered Local 545 and the contractors to cease and desist the practice of charging license fees to members of other locals and nonunion members. However, unlike the initial order, the owners of the companies and Local 545 were not required to pay back any of the fees collected from former Seaway workers. As the order was finalized in 1960, after the St. Lawrence Seaway and Power Project had been completed, it served more as a warning to union leaders and to Morrison and Knudsen, Walsh Construction, and B. Perini to not undertake similar hiring practices on future construction sites.[15]

Lowell Fitzsimmons

I was a dragline operator on the St. Lawrence Seaway dredging project between Massena and Alexandria Bay. Local 545 held just as much clout with the contractors as Local 322, if not more, since their members were highly trained and therefore harder to replace. One day a man from Texas came on the job and wanted to work days and make me work nights. I

16. Two night-shift workers stand under the legs of a three-hundred-ton gantry crane at the main power dam. Courtesy of Alfred Mellett and PASNY.

had quite a bit of influence with the union because they knew I was a good worker. So I called the union's business agent and I told him to come down to the job because I wanted to talk to him. We talked the situation over and I told him that I had a new superintendent, who had only been working there a few days and probably could be scared easily. The business agent told him that if I had to work nights he would shut the job down until I was back on days. I waited only a few minutes and the superintendent told me that I never had to worry about working nights.

Ray Singleton

I have a lifetime membership in the Operating Engineers union. I was born and raised in Conway, South Carolina, near Myrtle Beach. My ancestors moved to the area before the American Revolution. During World War II, I started working in the Charleston navy yards, when I was fifteen years old. I quit school and was interested in becoming an electrician and they hired me. I worked there for a year and a half. Then after the war was over there were so many unemployed electricians, I couldn't find a job. So I went to Florida with my dad, who was a crane and heavy equipment operator, to dig canals with a dragline and to learn how to run heavy equipment. In 1947 I operated a gantry crane for the first time on the Clarksville Dam in North Carolina. I didn't run any more of those until the Seaway started. At that time there were so many dams and locks being constructed that experienced equipment operators were in great demand. My brothers, my father, and I were some of the few who could handle all types of equipment, so we had no problem getting work.

I got out of the air force in October 1954 and for a year helped build the first Ohio Turnpike. I arrived in Massena in 1955 when the Seaway project was just getting started. I helped build the first bridge over the Grasse River with American Bridge and Steel Company. I worked on the Long Sault Dam from the beginning, and then later on the Barnhart Island Dam. Eventually my father and three brothers came to Massena to operate heavy equipment on the various project sites. I am the oldest. I was twenty-seven when I started working on the Seaway. I worked for Walsh Construction and then B. Perini. All the sites on the power dam and Seaway were bid as joint ventures, among Walsh, B. Perini, the Utah Construction, and Morrison-Knudsen. Those companies at that time probably did not individually have the equipment or the know-how to do all aspects of the construction of any of the sites. Walsh Construction had worked on the Clarksville Dam.

When I worked on the Clarksville Dam, there were a lot of old gantry crane operators, including Mr. Kilpatrick, who was almost eighty years old, and others who had constructed the Grand Coulee and some of the other big dams. I worked with a few of them again on the Long Sault Dam. If you didn't have experience as an equipment operator back then, it was almost impossible to get work. It was not a cut-and-dry situation to learn how to run that stuff. It was at least a three-year apprenticeship program to be qualified to operate heavy equipment safely.

The St. Lawrence Seaway and Power Project was constructed exclusively by union members. If the union hiring agents had problems locating operating engineers, they just called other locals. There were people from all over the United States who loved big construction jobs like power houses and dams. Some of the guys had worked on the Ohsway Dam in India, while a lot of the southerners were former TVA employees who had built hydrodams in the South. Those engineers and operators who had previous experience on that type of construction were vital to the various contractors on the Seaway. Carpenters and the other skilled workers who had worked on big power houses were recruited by the contractors from their locals across the country because they had experience. Most of the people in Massena were not skilled tradesmen, and therefore did not have the know-how to do the work. However, contractors hired local farmers to drive the large tractors, but not the big shovels, excavation equipment, or the gantry cranes.

Ontario Hydro officials took a different approach to labor relations. In 1955 Dr. Richard Hearn, chairman of Ontario Hydro, and G. Russell Harvey, chairman of the Allied Construction Council, announced the extension of the labor pact from the Sir Adam Beck Niagara Generating Station Number 2 on the Niagara Falls project with the Allied Construction Council, which represented seventeen unions. This document covered wages and working conditions, and formed a safety committee to look into the cause of accidents and ways to improve overall performance. Taking a cooperative approach led to smooth relationships among workers, supervisors, and Ontario Hydro and their contractors.[16]

Bill Goodrich and his assistant set up an office called the Central Employment Bureau within the Cornwall National Employment Service (NES) office. Heavy construction workers applied for jobs at their hometown NES office so they could wait to move to Cornwall until a position became available. At the onset of construction numerous workers migrated to the area and found no work because many long-term Hydro employees had transferred from the Niagara Falls project to work with Iroquois Constructors. By Inundation Day 22,420 clerks, operators, tradesmen, and laborers had been hired through the NES.[17]

Bill Goodrich

I was twenty-four when I graduated from McGill with a degree in industrial relations. I was employed by the Ontario Hydro personnel department to hire workers and deal with all aspects of employee relations on the Niagara Falls project. I was the third of fifteen Ontario Hydro employees to arrive in Cornwall from Niagara Falls. As the personnel officer, I hired 18,000 people to work on the Canadian side of the power dam. I recruited laborers from as far away as Kingston and Belleville, Ontario. I did all the hiring for the contractors plus the industrial relations aspect of it and all the bargaining.

When I needed skilled labor, I traveled to Montreal. Even though the unemployment rate in Cornwall when I arrived in 1954 was 18 percent and many of the local residents were on welfare, they were not interested in working for a living. In July 1956 there was an acute shortage of skilled workers on the Robert Saunders Dam, which threatened to delay the construction schedule. I put out an appeal to current employees to supply names of their former work mates and supervisors and encourage them to report to the nearest Employment Service Office and fill out an Ontario Hydro application.

I had a pretty superb relationship with the unions. The reason it was so superb was that I had dealt with them in Niagara Falls. About 50 percent of the electricians, operating engineers, and carpenters had been employed by Ontario Hydro on the Niagara Falls project. Prior to Robert Saunders Dam project, Ontario Hydro had done all of their own construction. This time the contractors completed all of the construction, but we handled everything for them from the aspect of personnel along with the National Employment Service Office (NES). The NES assigned two men to accept the applications and initially screened each individual. One chap was extremely good at locating and evaluating operating engineers. The other chap handled the recruiting of electricians and carpenters. If the local labor market could not fill the spots, they recruited other workers through the NES offices across the country. These two men would then send these individuals to my office where my friend and I would make the final hiring decisions.

Cyril Dumond

I was hired through the pipe fitters' union. There were fifty of us that worked two shifts. Our local wasn't big at the time. It was just starting up and we didn't have the numbers to supply the job, so we got workers from Toronto, Montreal, and from towns all over the province of Quebec.

We called up the leaders of other locals and asked them to send men to Cornwall. The people that worked on the power dam stayed on the job for its entirety.

I was a pipe-fitting supervisor on the power dam site for three years and ten months. Initially the pipe-fitting union brought in a superintendent from Toronto who answered to the subcontractor. I had a crew of men whom I supervised and inspected their work. I had counterparts among the electricians, carpenters, and masons who had the same responsibilities. Our superintendent didn't last very long and his successors didn't either. We went through quite a few of them. They would quit and go home because they didn't like to travel and be away from their families. None of my co-workers ever complained about Cornwall itself or the project. Many were just homesick. Some brought their families with them, while others worked Monday through Thursday and left on Friday. I liked working on the power dam because it was a multiyear job and was never boring. There were rod men, carpenters, plumbers, electricians, members of all the different trades who worked on the power dam. It was big deal for me because I had never built a dam before. It was a good experience.

Glenn Dafoe

I served as a union and shop steward for the carpenter's union. Once the number of workers increased, the unions were more involved and tradesmen could not do any tasks outside their specific job descriptions. I can recall two incidents when this impacted my work. In one of the houses we had moved, the electricians were waiting to put the service in for an electrical panel. On this project, the board that the panel was attached to had to be installed by the carpenters, but none of us were available, so they had to sit around and wait a few hours. On a normal construction job, the electricians would just nail the board to the wall panel and finish their wiring. Another time, one of my co-workers, who had actually been my employer before I had started working for Ontario Hydro, needed a power saw and we didn't have one at the house we were working on. He drove a station wagon and carried all of his own tools in the back. So this guy said he was going to use his power saw. I told him not to use it until I read a copy of the collective agreement to see if this was allowed. I asked my business agent two or three times to get a copy of the agreement, and he never gave it to me. Finally, I let him use his saw. It wasn't ten minutes until one of the electricians' stewards came over and wanted to tear a strip off me. He said he would have my job over that. I told him that I had asked my union for proof that this was against the rules, and I had not gotten a response. The next morning, the international representative of the carpenter's union from Toronto, Red Ogden, came to the house I was working on and told me that I couldn't allow my men to use their own tools. I replied that I intended to use whatever tools I could find, until someone provided me with a written document that stated otherwise. He told me I should take his word for it and we had a rather spirited discussion. This was one of the things that I detested about the union representatives. They wanted you to do as they said and not ask questions. I think what it boiled down to in my opinion was common sense.

My business agent paid me a visit the next day and asked me to have my man put his saw away until we got the situation straightened out. Within an hour, I had the collective agreement. One of the things that was in the agreement that the unions had with Ontario Hydro was that no tradesmen was allowed to provide any of his own power tools. They had to be supplied by the employer.

Joseph Couture

The union stewards always came to me with demands, but there was one I never forgot. They were threatening a wildcat strike, so the other foreman and I had a meeting in our office. I told them that our best bet was to fire everyone and then hire the men back who would work without the union. I told my boss I was going up to the highest part of the dam to hand out blue slips. He was worried that someone might try to push me off. I replied that if I went down I would take at least one other man with me. When I got to the top of the dam I told them if they didn't want to work through the strike, I would fire them. The first guy I asked said that he didn't want to defy the strike, so I started to write him a slip and then he asked if he could change his mind. No one else asked. Most of the time when the union stewards made a demand, I would have a meeting with them and tell them what I had to offer. Other times the requests were so crazy, I just refused on the spot. One of the most typical grievances was that sometimes the carpenters had to buy a certain tool to complete a section of the dam that the company didn't supply and the workers and the union would protest and want the contractor to pay for it.

4

Life on the Job

THE DAILY LIFE of a Seaway or power dam worker was long, challenging, and sometimes dangerous. The working conditions on the St. Lawrence Seaway and Power Project have been described by workers as brutal and unforgiving. Workers and contractors toiled in mud, water, and extreme temperatures atop high concrete walls to complete the most ambitious waterway and power plant ever attempted. Laborers, carpenters, and members of concrete crews began their day by reporting for one of three eight-hour shifts to pour concrete, build forms and house foundations, or transport gravel and other supplies. During this time span each dealt with large equipment, poor weather conditions, and a hazardous working environment littered with debris and populated by men from all trades. After their shift ended, many went to one of the local bars for a drink, and then returned home to catch a few hours of sleep before starting the whole process again.

Contractors strongly encouraged their men to work extra hours, because there was always a deadline that needed to be met. All levels of workers from engineers to laborers often worked overtime to review new plans, or to complete a concrete pour or painting job. A June 1958 Ontario Hydro accident report revealed that 22,967,424 hours had been logged on the Canadian side of the power dam. This intense schedule caused workers to be less alert and cautious around large machines and at extreme heights. Tragically these circumstances resulted in the deaths and injury of dozens from a fall, electrocution, drowning, or improper use of unfamiliar equipment. In the early days of the project, according to Jacques Lesstrang, "it became apparent to the people of both the United States and Canada that without international cooperation—not between

126

governments, but between men, sick, tired, covered with marine clay and concrete dust, men bruised and cursing and cold—the project could not have been completed on time, if indeed at all."[1]

The weather made the day-to-day lives of engineers, tradesmen, and laborers difficult. At the beginning of the project workers attempted to complete concrete lock walls and dredge canals even when faced with frozen ground and water. However, some conditions made these prospects impossible. In the winter of 1955 floating river ice interfered with tugboats and equipment being used on the construction of the cofferdam, and jams caused high water levels. As the project continued, contractors realized the impact of subzero temperatures on concrete integrity after inspectors deemed sections of the Eisenhower Lock inadequate. Based on this discovery, Corps and PASNY officials halted the pouring of concrete on the American side for three months every winter. Dragline operators also could not break through the thick surface ice and delayed their operations until spring. In the spring of 1956 mild temperatures and heavy rain caused a sudden thaw and muddy conditions, making roads slippery for truck drivers and work sites soggy messes for masons and laborers. On December 4, 1956, 175 laborers, some of them suffering from frostbite, walked off the Barnhart Island site when the mercury dropped to twenty-three degrees below zero. For two years these men had braved heavy wind and bone-chilling temperatures without adequate clothing or shelter. They had been pushed to their limits and refused to continue to risk losing a finger or limb to meet an impending deadline. In February 1958 PASNY reported numerous blizzards that threatened to interfere with the scheduled completion date. After three years men and machines broke down and contractors lost many supervisors and skilled workers who, anticipating the end of the project, had moved on to other construction sites. Uhl, Hall, and Rich, the Corps, and their contractors offered their remaining employees pay raises as incentives to delay their exodus until 1959.

As the local representatives of their agencies, the engineers who worked for Uhl, Hall, and Rich, the Corps, and Ontario Hydro struggled along with the contractors to deal with design and cost changes and daily dilemmas that arose. Engineers on all major construction projects face change orders. However, in hindsight the higher than normal amount on the Seaway and Power Project can be traced to the fact that the Corps of Engineers did not complete detailed plans of any of the locks, dams, or dredging projects before the contractor bid for each site. Instead, planners completed three thousand vague drawings for the Barnhart Island Power Dam, fifteen hundred for the Long Sault Dam, and more than eighty pages of general descriptions of materials to be dredged to accompany each bid. The Corps of Engineers and Uhl, Hall, and Rich left many specific design elements to be determined by the contractors because they had limited time between the passage of the Seaway bills and the bidding of the jobs. Because of the complexity and the unknown territory this project entered, the complaints overran the claims departments of Uhl, Hall, and Rich and the Corps. Each one required a comparison with the original contract and drawings and the approval of extra funds. As engineers had broad-based responsibilities, they worked more hours than the construction workers and played an important role on the project. Because they were salaried employees and not represented by a union, they stayed at work until a task was completed.

Jim Cotter

I worked in the main office of Uhl, Hall, and Rich with my partner, Don Wiles, receiving and transmitting drawings from our office in Boston to the appropriate contractors and the International Joint Board. Construction drawings by the hundreds came into our Massena office. Don Wiles and I wrote thousands of letters of direction to the appropriate contractors. We set up our engineering shop in the old Grange Hall and bided our time until the corrugated metal office building was completed near the future Long Sault and Barnhart Island structures. There was no honeymoon on the job as many construction contracts had already been awarded and additional ones were being awarded on a daily basis. The contracts covered all conceivable vital hardware and materials as well as the design of the actual locks and dams including all of the electrical wire, transformers, the mated turbines, generators, steel girders, railings, sand, gravel, and cement. Any price increases and changes went through our office. What made claims so hard was that underbidding was part of the game and a clever manipulator of facts could come to different conclusions entirely on any change order.

My partner and I had to go to the Power Authority to get more money for any redesigns and the settlement of any claims. We could make small design changes without their approval, and we just advised them that a change was being made. The Power Authority was not staffed to deal with the great engineering issues. They had their resident engineer, but he was not there to approve things of that nature. They did not know what was going on. But they wanted to act like they had control over everything. Don and I had to respect their directives as they were the money source.

The basic meat and potatoes for contractors from an engineering point are the drawings that show every detail of what work needed to be done. Once a week Dan and I received these rolls of drawings from our Boston office that weighed forty pounds and were eighteen inches in diameter. We had to sort through all of them and make sure the right ones went to the right contractor and the International Joint Board got their copy so they could review everything. But before writing these letters of directive, we had to review these drawings to see if there were any changes that had occurred from the prior edition. If changes had occurred, I was expected to acknowledge them and ask the contractor to give me a quote or some resolution of any problem that might arise due to these alterations and not allow problems to accumulate to the end of the contract. I saw changes down to the very time I left. It had reached the point on the Barnhart Island Dam that the architects made these little tiny changes. I said, "My God why make a little change like that regarding a little strip of metal up the railing of a stairway?" All those things resulted in claims. Many of these alterations were not motivated by construction or safety concerns, but for aesthetic reasons.

Joe Marmo

As an office engineer, I did everything that was needed for the entire project, which was spread over forty miles. I didn't concentrate on any one thing in particular. At the beginning, I was the only office engineer. My hours were set by Uhl, Hall, and Rich. I was scheduled for five eight-hour days and four hours on Saturday. If anything else came up, I just put in the needed hours. I knew what I had to do, and had pretty steady hours. I started out with the Bureau of Reclamation making about $3,100 and then I got about $5,000 on the power project when Uhl, Hall, and Rich increased salaries to keep people up there like me as the project wound down. Every nine months I stayed I got an extra $3,000 a year. By the time I left in 1960, I was earning $13,000.

In claims, Roy Simonds was my boss. We evaluated all the claims together. He was very highly organized and a very intelligent fellow. Roy was the one who got me into the claims section. He asked if they would transfer me in, so we could work together. Whenever a claim was submitted to our office, it stood four feet high and contained about fourteen books of information. A contractor loves a change because he makes a 25 percent profit on the change. There is no competition so he has to convince the overseeing authority that the price is right. If they don't receive what they feel is fair compensation, they cry bloody murder for the rest of the project. Every time we had one submitted, Jay Goodstein, a lawyer for PASNY, wanted us to find the original drawings to compare to the newer version. Roy and I pumped Mr. Goldstein with information and gave him our recommendations. He then took this data and negotiated with the Perini Corporation. All the contractors submitted claims, but it was only Perini who was paid because he had the biggest share of the Barnhart Power plant. He would then settle the claims with the subcontractors under him. Perini was the one who had the power to stop the job and take PASNY to court.

Roy Simonds

I was in charge of addressing all of the change orders for Uhl, Hall, and Rich. It was the most interesting job in my career, so I worked six days a week and sometimes a seventh day to make sure the work got done. I didn't mind working overtime. Every day I came to the office, I was dealing with something unique. There were a lot of problems to solve. It kept me on my toes all the time. There was so much activity and so much variation because every type of construction you can think of was all happening at the same time. The dredging and the concrete pouring went on twenty-four hours a day, seven days a week. Uhl, Hall, and Rich moved up the schedule considerably because PASNY administrators, particularly

Robert Moses, realized how much additional money he could get from the power production if we finished early, so they accelerated a lot of contracts and paid dearly for it. The way PASNY got the power aspect justified was that they promised to finish in a certain amount of time, so administrators decided to go ahead with it and take care of things as they went.

On the building of the power house and the power dam, Robert Moses decided he wanted to make that a lot fancier than originally designed. He made an awful lot of changes. We directed the contractor to do the work and then compensated them for it after the fact. That put a financial burden on the contractor. We had to make many of them redo work, but that is typical on any construction job.

Generally, I had some kind of meeting with one of our various contractors every day. We discussed their work and changes or repairs that needed to be made based on our inspections. The four largest contractors in the world, Morrison-Knudsen, Peter Kiewit, A. J. Walsh, and B. Perini, jointly ventured most of the projects on the Seaway and Power Project. Otherwise, they couldn't have handled it. These companies were so large they could handle the extra financial costs better than some of the smaller ones. Each sponsored a different project. The contractors would bring me design problems as they encountered them on the job. If the amount of concrete or materials they needed to use differed from what was included in their bid or the measurements varied from what was drawn on the plans, we would try to find a resolution together. We had one or two contractors that had to be removed from their contracts, but they were small. The large contractors seemed to have less of a hassle because they had the experience and workers to deal with redoing a particular segment of the project or weather-related delays. Sometimes they had to put on additional people to meet an impending deadline or make up for lost time after a long cold spell. The labor force in Massena was often strained when all those projects were being constructed at the same time, but the larger contractors just brought in workers from their other sites.

Jack Bryant
I got in the habit of working long hours. As an engineer, I got paid by the month, not by the hour. I usually worked all day Saturday. Uhl, Hall, and

Rich gave me a car, which was huge plum because I missed the carpool times. It was really quite a dedicated bunch. Most of us from the Bureau of Reclamation were experienced in administering the contracts and inspecting the work. The others were mostly from TVA and their system of construction was quite different. They expected that if they saw problems, they would solve it themselves without putting in a change order. But you couldn't do that in construction by contract.

I remember I got in a fight with some of the TVA people because early on they said that contractors were all honorable people and they would do what they were supposed to. If something cost extra money to do, that was part of the game. I tried to tell them that they could not do that in contract construction. Anything that deviated from the original cost or design had to be approved by the Power Authority. The contract sets the rules and the quantity estimates, but almost always the quantities will go up or down and quite often the quality of the work and the nature of the work will change because of subsurface conditions or some other damn thing.

There were an unusually large number of change orders compared with other jobs I had worked on. I and my co-workers analyzed each contractor's complaint and determined whether we agreed at the project level that he had a legitimate beef. These claims related to design changes, or unforeseen subsurface conditions, or sometimes just plain underbidding. There were clauses in each contract that provided for price adjustments and time adjustments, if there were subsurface conditions materially different from those indicated or if they were more than anticipated. That was usually the basis of their claim and that was certainly the basis for those who were paid. A lot of them weren't paid. Some of the contractors on the power plant got in such trouble at such a high level that some sort of accommodation was made for them way above us at the Moses level.

I attended a couple of meetings in New York City where Lou Perini, the head of B. Perini and Sons, was making a pitch to the Power Authority hierarchy for compensation for changes and cost overages. It is unquestionably true that many of the contractors lost a lot of money. Whether they were entitled to be made whole again became a question that was decided by Robert Moses. What we could justify from an engineering point of view we did. We didn't give final answers; we just offered recommendations and most of the time the recommendations were followed.

The various cement and electrical inspectors assumed the large responsibility of ensuring the quality of the work being done at each location as well as adherence to the tight time schedules. Often these men oversaw several concrete pours or electrical work being completed simultaneously. The engineers relied on this fleet of inspectors to guarantee that contractors properly executed the official plans and did not cut corners at the various locks, dams, and dredging sites. Inspectors had the authority to shut down a concrete pour or wiring project if they deemed the work to be faulty. Many were much younger than the contractor's supervisors, but gained their respect based on the backing of the project managers and their positions. On the American side Uhl, Hall, and Rich and the Corps tried to inspect their sites with the lowest number of men possible based on financial constraints. Many concrete pours and excavations took place with little inspection by the supervising authorities, which allowed for substandard construction techniques to go unnoticed. In contrast, Ontario Hydro inspectors completed training programs and faced surprise visits from Hydro administrators. At the beginning of each shift, foremen charged their inspectors with observing a very specific section of a concrete pour or carpentry project, guaranteeing the constant supervision of contractors and their workforce.

Ambrose Andre

At the beginning of the Eisenhower Lock project, I was a drilling and grouting inspector. The Corps had a grouting program that called for a curtain wall around the entire perimeter of the chamber. The holes were only a few feet apart and each hole was grouted one hundred feet below the surface of the rock. Somewhere along the way they decided it was too expensive. Workers grouted the upper sill, but the chamber grouting was eliminated. I think it was a cost issue. This wasn't holding up the operation. Concrete crews were pouring concrete at the same time. The grouting could have continued for another six months and not interfered with anything. Every monolith had these holes and our instructions were just to put grout in and fill the holes up so that we wouldn't have these open holes that would let water flow from one stratum to the next. Sometimes we would put grout in one hole and it would come out of the rock one hundred feet away. We had one hole that boiled water about a foot and a half in the air. If you put your foot on that hole then another one would pop up. It was definitely interconnected. The contractor put ten thousand bags of cement in the grout and got five dollars a bag. Back then that was a lot of money. On the bid document their estimated number of bags of grout was one hundred thousand and we actually put in ten thousand. It was one of those things that the government reserved the right to delete part or all of it.

The Corps was also short of concrete inspectors because they wanted to make do with as few people as possible. I still resent that to this day. I graduated from Clarkson University in the summer of 1955 and in the summer of 1956 I became a concrete inspector with absolutely no training. As a graduate engineer I was supposed to know everything about design and construction. During my shift, I was the only government person on the job. There were three pours going on simultaneously. There was back-fill being placed, forms were being built, and the water stopping was being welded. My normal job was form inspection. I made sure that the forms were where they should be and the embedded metals were in place and properly supported. Initially there were three concrete inspectors, including myself. However, one fell off a fifteen-foot form and broke his leg. Another fellow almost cut off three of his fingers when they got pinched in a concrete bucket. The Corps employment office only hired one replacement, which meant there were only two of us to cover a twenty-four-hour time period. Since we worked ten-hour shifts, this left from ten until two in the afternoon with no concrete inspectors. The crews were pouring without supervision. Being my first big construction job, I didn't know if this was normal or not. I wasn't comfortable with it, but it became the norm.

Bill Spriggs

My main responsibility for the duration of the project was as an electrical inspector. There should have been more than one electrical inspector on each shift. It was impossible for me to keep track of all the work going on during my shift. I couldn't even make it from one end of the project to another in eight hours. A few times, I found a problem with the wiring because the contractor had not followed the plans. So I shut him down. I didn't do that very often because it cost plenty of dollars and put the work behind schedule. The Corps was awfully cheap. They didn't mind taking advantage of me, by paying me to do inspections when I had no experience.

I worked seven days, sixteen hours a day, and any other hours that were necessary. The reason I had to work these hours was that I was the only one who did my job. I was interested in what I was doing and the challenge was there. I am of the old school, the Depression work ethic that you work until the job is done. I didn't do just my job; I also worked with the soils, the structural, and the mechanical inspectors. In fact we put up a new building in town and I was made the inspector on it because the man who was supposed to do it was a young lieutenant who went on leave for thirty days and never came back. You had to be bold and be willing to say, "I can do that." I even worked with the well drillers. They were doing core boring at the lock's base. The Corps bore down about seventy feet and recovered soil samples and found the voids that needed to be grouted so the lock walls were constructed on a stable platform. They hit these big geysers that were powered by the river flow that were sometimes five to six feet high. They piped in 3,800 bags of grout and you never could find it. It never did do anything. It just went down a rat hole. They did it on the power dam as well.

I also worked closely with the electrical workers' union stewards and their members on the job site. Most of them understood the stiff time-table we were under to complete the work by Inundation Day, so they didn't argue when I pushed them to work harder and faster. They had a few members who complained that we were working them too hard. But everyone was being paid well and seemed to be enjoying what they were doing, so I had little resistance.

Donald Rankin

On a typical day Ontario Hydro inspectors kept track of the material and concrete quantities and our survey people were checking that the work was being done according to the plans and was in the right place. In the project manager's office, my staff and I would receive all the drawings and hour by hour we would determine as far as we were able that the drawings were being followed. I was primarily concerned that the correct amount of steel reinforcing was done and that the safety and construction codes were being adhered to. I also dealt with Iroquois Constructors foremen and superintendents on matters related to their responsibility and to see that they were doing their jobs properly. The chief function of the owners' engineering group was to act as inspectors, and of course there were quite a considerable number of projects that were going on during the dam construction aside from pouring concrete and setting reinforcement steel. Most of the time it was just a day-to-day involvement to see as far as possible that the work was carried out properly. Nothing is ever performed perfectly, but I suppose you endeavor to obtain a result that is as near to satisfactory as possible.

Arnold Shane

I was one of fifty-six inspectors working for Ontario Hydro on the power dam site. A few inspectors were assigned to the mixing lab testing the various aggregates, while others were in the office tracking the pours, and the remainder, like myself, were on-site. As an inspector, I was scheduled for one of the three eight-hour shifts. It was my job to inspect the one pour that I was assigned to. Some days there were as many as twelve pours all going on at the same time, so there would be twelve inspectors on hand, one for each pour. You can't just have one inspector on-site watching all of them at the same time because the pours were happening on three levels, sometimes forty feet apart. I had to ensure that the proper mixture was used, that it was adequately vibrated and did not contain too much air that could cause honeycombing. I also had to make sure that we were not pouring too quickly and overstressing the forms. When that happened carpenters were called to make repairs, while the concrete crews moved to another section of the pour. Our contractors poured concrete around the clock. Their workers would start on Sunday and pour straight through until the next Saturday. They used all different mixes for different pours.

In the winter the carpenters got the forms ready in a protected area that was tarped off and heated with "salamanders." The poured concrete was also kept heated until it set. Then the tarps were removed and carried to the next pour location.

Concrete pouring comprised the largest component of the construction of locks and dams on the Seaway and Power Project. Crews placed over 887,000 yards on the Robert Moses–Robert H. Saunders Power Dam alone. The size of the pours on the American and Canadian side differed as did the mixture of the concrete. Researchers and engineers in the concrete plants at the various sites determined the chemicals that needed to be added to prevent cracking and freezing in the winter, and carpenters and laborers tarped off and heated their work areas. The process of transporting the material to the batch plants, mixing the concrete, and finally pouring the material into the forms required the cooperation of a variety of trades. Truckers brought in gravel from off-site pits to the concrete plants on-site. The plant operators in turn took the gravel and mixed it with the right amount of sand and water for the consistency and strength required for the particular pours that day. Truckers then transported the mixture in their truck beds in big buckets to crane operators who lifted the material to vibrating crews who shook out the bubbles and dispensed the concrete into forms set by carpenters. American contractors halted their concrete pouring for three months in the winter, while Ontario Hydro continued their operations year-round. They had mastered the art of heating and pouring concrete on projects in northern Canada.

David Flewelling

The normal workday was eight hours; however, in the period that I worked on Grasse River Lock, I normally worked ten, and I sometimes averaged about twelve hours a day. Some weeks I put in eighty hours. There were a couple of weeks prior to the start of the summer season when the drafting room was really busy. For the most part I worked six days and had Sunday off. I was making $2.12 an hour. That was double the wages on any other job.

Each concrete pour had a guide, which showed the field people the size and required elements. My job was to outline each lift and illustrate the location of the rebar and indents. I had taken a drafting class at college, but I was basically trained on the job. It wasn't brain surgery. I could read plans and I had a high school and college background and I had taken a year and a half of chemical engineering courses. It was mostly civil engineering. The project engineer had the timeline for each of the sites and our

17. At the Long Sault Dam in March 1958, a forty-ton caisson is being lowered from the deck into position on spillway bay two by a gantry crane operator. Courtesy of Alfred Mellett and PASNY.

drawings had to be completed four to six weeks ahead of the scheduled commencement of that part of the construction. Once we produced the drawings, Uhl, Hall, and Rich engineers reviewed them and added the necessary corrections before they were sent out to the field. Then a record drawing was sent to the project engineer.

Ray Singleton

Once the contractors started pouring concrete, they didn't stop until the form was filled. When I was operating a gantry crane, which was used around the clock, if the operator on the next shift didn't come in, I worked a double shift. Other times, my supervisor asked me to split a twenty-four-hour shift with another guy. They started at the bottom of both dams with fifteen-yard pours that extended all the way out to the spillways. As they got closer to the top of the structure, they got shorter. The concrete workers poured the concrete in staggered blocks, like a checkerboard. The carpenters built the forms and then me and my fellow crane operators put them in place along with the steel, pipes, and rebar that reinforced and drained each structure. As a crane operator, I was pretty much involved in the whole operation.

Thomas Sherry

The construction went on twenty-four hours a day. I worked a ten-hour shift from 7:00 A.M. to 4:30 P.M. Another shift come right in and worked under the lights all night long. It was continuous. The contractors were in a big hurry to get the job done. I liked to keep busy, so I was mad when the foreman would tell me to go and hide for the rest of the shift in the tunnel because there was no more patching work for me to do. It didn't happen every day. Sometimes it was due to a lack of materials or concrete pours not being completed. I don't know if the contractors were getting paid based on how many people were on the job. A lot of that kind of crooked stuff was going on. I would go crazy because I was used to working all the time.

During my first two-month stint on the Grasse River Lock, I worked for Morrison-Knudsen. They were one of the biggest construction companies worldwide. The carpenters put together square concrete forms by bolting together four-by-eights. The crane operators would put these forms in place; workers then poured the concrete and waited for it to set. When the concrete had cured and the crane operators removed the forms, there were holes underneath made by the expansion bolts that held the forms in place. It was my job to remove these bolts and fill in the holes with mortar. I got fifty cents more an hour for doing this dangerous work in a bosun chair. I had never been up on a chair like that. It was like sitting on a swing. It was a little wooden chair with a rope and a pulley. I had to pull myself up about fifty feet off the ground and tie a knot in the rope and then I did my work. If I didn't have the knot tied right, I was going for a long ride down. It was pretty scary. I learned how to use the chair from masons who had worked on other big dam projects in the West. On other sections where I needed to patch, I had to climb up the lock wall. If it was a sloped wall, I put on a harness like a paratrooper and walked up and did some patch work. It was like walking up the side of a roof. I didn't mind that. I had to keep walking and the pulley kept me moving forward. Then I would stop, tie a knot and do my work, and then move ahead some more.

There was a lot of rebar sticking up out of the ground like a tiger trap. If I ever hit that I was dead. I didn't care for that much, but I was young and I figured I would take the risk and make the extra fifty cents more an hour. I probably would never do it again. Back then I was filled with the spirit of adventure.

David Manley

When I got promoted to the concrete crew, I often went in at midnight and stood around a lot. We wouldn't get the first bucket of concrete until two in the morning. The men in the plant had to mix the cement and get the crane operators to lift it to the pour location. When the cement started coming, however, we had to work on that pour until it was done. If someone had to go to the bathroom, I had to take his spot. Vibrating the concrete was very difficult because the drum weighed ninety pounds. That is why the young people did it. My crew and I filled the drums with concrete, vibrated the mixture for the right amount of time, and then dumped it into the forms and smoothed it out. I had to be careful because if I didn't move my feet, they got stuck in the setting concrete. Sometimes the guys played a joke on me and moved the vibrator and poured wet concrete behind me when

18. An October 19, 1956, photo of the St. Lawrence Power Dam illustrates the rebar installed to strengthen the concrete, the tracks for the lift gates, and the gantry cranes for construction and gate operation. Courtesy of Alfred Mellett and PASNY.

I wasn't looking so my feet would get stuck. I also had to wear a raincoat even when it wasn't raining because there was so much water around.

I worked six days a week pouring concrete and then on the seventh day I cleaned the buckets and all the tools. I also worked every other Sunday. Once my crew and I finished our pour, there was very little time before we removed the form and someone was green cutting the concrete and watering it down. Men were also sandblasting and trying to put the forms up simultaneously in a very small area. There was quite a time constraint.

Roderick Nicklaw

In the summer of 1955, I worked at Robinson's Bay Lock for Morrison-Knudsen where the tunnel goes underneath the structure. It is called the Eisenhower Lock now. I always remember it because Dave Manley, myself, and several other guys that we knew vibrated concrete on a crew on the midnight shift. We made $2.17 per hour and that was big dollars back then. If you made a buck an hour back then, you were making good money. We had a four-man crew and a boss. The concrete plant workers would send up these buckets on a crane and then the operator would open up the mouth of the bucket and pour the material into the vibrator. Then two guys vibrated the concrete to settle it and get out the air pockets, and then the other two guys would dump the bucket. We called them bucket busters. It was a tough job. They were ninety-pound vibrators and I only weighed 160 pounds. Two guys would run the vibrator for an hour, and then the next two guys would run it because if you had to run them for eight hours yourself, it was a little bit much. Sometimes we had to wait for the concrete truck or the cranes got held up, so we got some breaks, but when everything was running right, I worked hard. Sometimes we worked the whole shift without a break.

The Indians worked on what they called the bull crew. They used to carry all these vibrators and hoses up to the section where we were going to pour. They climbed up makeshift ladders on the side of the lock made of two-by-fours. Safety was not like it is today. It was risky work. It was dark, cold, and slippery.

In the summertime when I was out of school, I worked the four o'clock shift for a while. Otherwise I worked the night shift. During the day, the carpenters prepared all the sections to be poured. The concrete workers cut all the green off the concrete. They removed any loose particles from the completed pours with a high-pressure hose. If the surface of the dried concrete was rough, the new batch of concrete that we poured on top wouldn't adhere to it. Carpenters oiled all the forms so the wet concrete wouldn't stick to them and placed them during the day, so we could pour at night. The on-site batch plant operators also determined the mixture they needed to make for that specific pour. By the time I arrived with my crew at night, everything was ready to immediately start pouring.

Some funny things happened every day. The contractors set up these portable outhouses. They would dig an eight-by-eight square hole and skid these huts over it. There was always a trail from the parking lot where they had dragged one of those crappers down to where we worked. After that hole got filled, they would just cover it up with dirt and skid the outhouse over another hole. Every once in a while, someone would walk into one of those holes that had just been covered up. It was a dirty job.

A messy situation also occurred when they first started a pour. The crane operators would hold a bucket of soupy grout six feet above the floor and then they would open up the bottom of the buckets, and all the stuff would come out. It was like a wave of soupy concrete. If you weren't looking, the guys would always try to put one in right behind you, so you would get covered with concrete. It would almost knock you over like a tidal wave. They would only do that to you once and then you would wise up and pay attention.

Arnold Shane

Ontario Hydro did a lot of research about concrete. I soon learned that it was much more than just cement. You had to design each mixture based on what you wanted to use it for. There are fifty different mixes that you can come up with to do certain things. First, I worked with concrete experts in the lab testing the aggregate. We then took several concrete cylinder samples at every pour on the power dam and let it cure, for seven, ten, and fourteen days. When the allotted time elapsed, we broke them to test their strength. Some we let rest for twenty-eight days. We buried cylinders in crates with ten different mixes used on pours at the power dam. They were to be dug up every ten years after the job was done to study how the curing continued underground and if the concrete maintained its strength over time.

Ontario Hydro also mixed all of its own concrete on each project location. The biggest concrete plant was at the Robert H. Saunders Dam. The dry cement was brought in on railroad cars and then pumped across to the storage silos through a pipe under the canal. Then when an order came in for concrete for one of the pours, the attendants would mix the dry cement with the appropriate amount of water, sand, and stone. At one time we were pumping it over to the Americans for use on their side of the dam when there was a cement strike. The concrete for the house foundations at the town sites was all brought in by truck from off-site batch plants.

John Moss

From 1955 to 1958 I supervised the concrete pours and the contractors' work on the power dam. I worked all three shifts and I worked twelve-hour shifts sometimes. It all depended on what work was scheduled. The night shift was the hardest. I didn't have any trouble with the contractors. They knew they had to conform to what I said, so they didn't question me at all. They tried to push me a little bit, or hurry me up, if they were anxious to get a pour started and I seemed to be holding them up. They didn't like that very much. I would be the sole inspector on one pour. There might be five or six pours going on at the same time. There were some of them that went on for days. In fact they never stopped pouring. There was a limit, however, to how much wet concrete you could have in a pour. Depending on the size and rate of the pour, there could be a six-foot head of liquid of concrete before you had to let it rest. If there wasn't any concrete being poured that day and they didn't want to lose any of the money, my boss sent me to oversee the concrete work on the new villages.

I sometimes went over and saw what the Americans were doing. I was always amazed to see the differences between our concrete pours. They constructed their side of the power dam with five-foot pours. Their concrete crews would start a pour and finish it in one shift, while ours might start one and end it days later. Mostly it was because the American contractors didn't have experience doing the large pours. I am sure they had not done a project like this before. They also didn't have the experience with winter concrete pouring. Ontario Hydro had developed different mixes and methods for pouring concrete in the cold. Most of the people I worked with were skilled and experienced, while the Americans were not. The contractors had problems getting men to work, especially in the cold weather because many of their employees were southerners.

Cyril Dumond

On the Canadian side of the power dam workers were scheduled on one of three eight-hour shifts. I worked until 4:30 P.M. and the other shift would come on to work until 12:30 A.M. My job was to fit all of the various drainage pipes that were used in different parts of the dam. There is a lot of drainage in concrete pours. Most of the machinery I used was run by

19. The St. Lawrence Power Dam looking from the American side to the Canadian side in November 1956. Courtesy of Alfred Mellett and PASNY.

air. My salary was based on the average rate all over Canada. At the time I was making $1.15 an hour.

I remember waiting around a lot in the cold. There were big pieces of pipes that were going into each of the pours. If the crane operators were too busy to get our pipes to us that needed to be fitted, we had to wait. Then we had to wait again for the crane to come back to assist us with

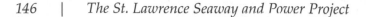

20. The same vantage point of the completed dam in 1958. Courtesy of Alfred Mellett and PASNY.

placing the pipes in the dam walls. Sometimes I made two pipe fitters and one welder work together when the crane was under our control, so we could get the most work done in the limited time we had. We had to get our piping in and do it fast. We also had to get all the pipes fitted and welded ahead of time so that the crane operator would just pick up our pipe and put it in the pour without a delay. Initially Ontario Hydro had only one crane on the main dam site that ran on tracks. So I had a hard time getting the crane because I was competing with the supervisors from all the other trades. The operators went back and forth on the dam all night lifting forms, pouring concrete, placing piping, and moving all of the other heavy material. Then Ontario Hydro assembled two ninety-ton gantry cranes. Once we got the gantry cranes, it took off a lot of pressure because the operators could go all over the span and there was less time-sharing. The mounted cranes could only lift sixty tons compared to the ones that were on the gantry that could lift up to ninety tons.

Carpenters on the Seaway and Power Project performed the most varied tasks. On the power dam and the lock sites, they built concrete forms and assisted in their placement on the ascending walls, which concrete crews then filled with concrete. After each section dried, they cleaned and inspected each form for damage and prepared it for the next pour. On the new town sites, once Ontario Hydro surveyors measured the houses to be moved, carpenters prepared them for transport and repaired the interior and exterior features upon their placement on their new foundations. Often their work entailed installing new roofs, building sidewalk forms, or patching walls. Many of these men had worked on other private and public works construction projects. Others learned their trade on the Seaway and Power Dam Project and continued to work in this profession in the private sector after 1958. Because the completion of carpenters' tasks was interrelated and typically delayed by the activities of other tradesmen, they frequently worked longer shifts and reaped the financial benefits.

Ira Miller

I worked on the Eisenhower Lock. I was put on the four o'clock shift for a while and then I went on the day shift. I was hauling milk when I was working on the four o'clock shift. I would deliver my milk and then go to work on the Seaway as a carpenter. I got four hours of sleep out of twenty-four. Basically it was the same as building houses. Everything had to be level and square. If I did that I didn't get into any trouble. Most days I built the forms and set them, and then tore them down and put them back together. The Corps inspectors would also check my work and complain to my supervisors if there was something wrong. They also were around on payday to check my paycheck and make sure that the contractors were

21. This aerial view of the Long Sault Dam in October 1957 shows the structure near completion. Roadway sections on top of the piers are being formed and concreted. Courtesy of Alfred Mellett and PASNY.

paying the right rate. I started out at $2.85 at Long Sault and finished a few years later making $3.85. I worked every day.

Joseph Couture

On the Cornwall dam project the men on the day shift would stay on if they had only worked a few hours during the day. We would keep them an extra hour or two to complete the work and get ready for the pour. People wanted to work and earn a living. I worked a lot of overtime to the point where I sometimes worked on Saturdays for twelve hours, which was time and a half and up to eighteen hours otherwise. We also got double time on Sunday. I worked 102 days without a day off. I would take a quick break and eat half a sandwich, but sometimes I would punch in at

8:30 A.M. and not eat until 4:30 P.M. My job kept me busy because I was in charge of a large area.

When I was younger the big projects were what I went after. Men were willing to work long hours back then to get the job done. They were good men and didn't mind the weather or the long hours. I could call ten men and get them to come in on Sunday and they would welcome it. You couldn't get someone to do that anymore. If we were well organized on Saturday and Sunday, we would have the equipment to ourselves to finish a pour. The concrete crews would have poured more concrete and worked faster if the contractors could have found more equipment. I had to plan my week around the two cranes we had.

I kept mostly the same people for the whole project. Hardly any of my carpenters quit. Most workers were willing to do a day's work and do what I told them. Some of the older men found it hard to accept a younger man in charge. When I was the general foreman, I had fourteen foremen under me. I was responsible for building and getting the forms ready and securing them to make sure the form would hold. The man responsible for the bottom part of the lock was a Polish guy who had been over here for a while and was twenty years older than me. We got along fine.

George Haineault

I worked on one of the three eight-hour shifts for Iroquois Constructors as a carpenter on the upstream side of the power dam. I rotated to the next shift every two weeks. Some men didn't want to give up their graveyard shift for a day one, if they were a farmer or working a second job. One of my first jobs was working on the forms for the ice chutes at the end of the dam. They were designed to allow for the blasting of ice if there was ever a big blockage. In order to do that job, I had to climb a ladder, which was one hundred feet in the air with a landing about every fifty feet. I was a young man then and wanted to show how well I could handle heights.

The concrete crews poured concrete around the clock. The pours were all staggered, so we always had one section to move onto. My fellow carpenters and I would put up the forms for one pour and when they were filling that one with concrete, we would place another one. We would watch each of our pours and the temperature of the cement as it was coming in because at the rate they were pouring it was quite a weight on the forms. Every once in a while if the bolts started to loosen, I had to go up and tighten a sleeve. These pours were so big that they would sometimes stop pouring one side of the wall and let it set and finish it a few hours or days later.

If I was busy during my shift, it was okay. But if I went there and sat around it was terrible because I had too much time to think about all the dangerous things that were going on around me. Every once in a while when a gantry got held up or broke down, we had to take a long break. Other times, the concrete plant would get backed up. On those nights, my boss would tell me to get lost and it was horrible. At three o'clock in the morning, I didn't think of the most pleasant things. I would look up and see all the dangers around me. I was so busy at other times and worked the whole shift without a break.

One cold night I was watching the cement pour and I put my hammer on the side of the form at 4:00 A.M. for several hours. It was as cold as hell. There were steel rods inside these forms. One of my jobs was to cover them with six-inch wing nuts, which I hit with my hammer. That night I missed the head of my hammer, and instead hit the side of the sleeve and it snapped in two. I sent it to the hammer maker and they sent me a new one.

William Rutley

I had two positions with Ontario Hydro. I surveyed houses and determined whether they should be moved to one of the two townships. I also worked in the warehouse to estimate the value of their content. I surveyed two or three houses a day. Then I would go into the office and drew them up. If I didn't think the house was worth moving, I didn't put a lot of time into it. When I arrived to survey a house, I told the owners who I was and why I was there and if there were any problems, I didn't do the survey. I would return later with a supervisor or the police. Then when I was assigned to the warehouse, some of the days the movers were bringing in the contents of more than one house a day. There was always lots of work to do. I had done office and surveying work in the mines in Sudbury, Quebec, so I had experience in that area. The surveyors relied on me to do a lot of the costing on each house.

In June 1957 I got laid off from Hydro and went to work for Atlas, a subcontractor on the power dam project, for six months. I worked on the night shift in an old warehouse ordering parts for bulldozers and the other equipment. If a piece of equipment broke down, it was my job to find out what was wrong with it, order the parts, and make sure it got fixed. The mechanics brought me the order number and I would leave it for the day man. If it was a special piece of equipment that was vital to the daily construction, I would have to get in contact with a superintendent and have him put in a request for an expedited order. We were not officially open twenty-four hours a day, but someone was always there in case of an emergency. The equipment operators worked a ten-hour day and in between shifts they checked over equipment. There were lots of break-downs when they started because the weather conditions and the twenty-four-hour work schedule were all new for some of the contractors. Many of them soon learned that if they operated a piece of equipment ten hours a day, six days a week, it wore out quickly.

When I was working in the mines I got $0.75 an hour. Hydro initially paid me $1.60 and then gave me a raise to $1.80, which was fairly good money then. Overtime was time and a half. I mostly worked five days a week. But if there was anything special, like a transport load coming, I would have to work on Saturday, but not on Sunday. There was no transport running on Sunday back then. It was supposed to be an eight-hour day, but I worked a lot of hours. There were lots of nights that I logged a double shift or an extra seven or eight hours.

Glenn Dafoe
Every day I came work, there was always something different to do. Some days, I worked on a house that had been put on a new foundation that needed to be shimmed up. The next day I was installing beam fillings, new basement windows, or basement steps. Other houses needed their porches rebuilt. We also poured sidewalks, erected scaffolding for the brick layers, and built outside chimneys for new furnaces. I did finish work, foundation work, and form work, whatever had to be done. That is all a part of being a general carpenter.

I was also involved in the filming of a television documentary by one of the Canadian networks about the house moving. The producer and the camera crew came to Iroquois to see how a house was moved and how quickly the services were hooked up at the new location. I worked the whole weekend to get the house they selected to film ready to move. On Monday, the mover arrived and lifted it up from the foundation, transported it out to the street, and covered up the tire tracks. There were two of us carpenters involved and they wanted us to sit on the mover as it transported the house from its original site to its new site to install the new beam filling—and I refused. It wasn't the way we normally did it. The documentary writer said that it only took forty-five minutes to move the house. It took fifteen minutes to lift the house off its foundation, fifteen minutes to move it to the new site, and fifteen minutes to have all the services hooked up. But they didn't say how much preparatory work had gone in beforehand and how much installation work took place after the house was set on its new foundation. For the filming, three linemen were waiting at the new house site to go up the pole to do the hookup and turn the power on. Several plumbers were there as well: one with a garden house to hook up the water for the drinking supply and another one to hook up the connection from the sewer until the permanent connections were made. It didn't actually happen that way on a day-to-day basis, but it made for good television. The whole process would often take several days if all of the tradesmen were tied up on another house.

The laborers and painters garnered the lowest wages of between $0.90 to $2.00 an hour. The laborers' varying pay scale reflected the variety of jobs that they performed on the project. For example, a laborer who carried bags of cement and shoveled dirt earned $0.90 an hour, while a laborer who serviced equipment would be paid $2.00 an hour. Others assisted welders or masons by providing them with tools or mortar. The knowledge and skill a laborer used every day on the job determined his salary. A laborer who undertook further training in order to learn new skills easily moved up the pay scale or acquired a job as a skilled worker or an equipment operator and received an enormous raise. A worker also increased his pay by working overtime.

David Manley

As a general laborer, I performed numerous menial tasks. First, I picked up lumber and stones. When they blasted, everything didn't go in a neat pile, so I was there to clean up. I also assisted the drivers of the big Euclids that came in loaded with stone to make the walls back up to the spot where they needed to dump. It was kind of a boring job. Finally, I cleaned the cement workers' and the finishers' tools. It was really dirty and everything was wet. There was mud all over the place.

Thomas Rink

I started with Uhl, Hall, and Rich digging test pits and testing soil density on the Barnhart Island dike. After that I was hired by Meritt-Chapman and Scott to assist a surveyor on the Long Sault Dam. Eventually I joined the laborers' union and was sent to the Eisenhower Lock in the winter with one other guy to tend fourteen masons. We supplied them with mortar and other materials to complete the brick and the tile on the control house. My final assignment was cutting trees for Stormont Construction, a Canadian outfit that cleared the land for the power pool. Hundreds of acres of woods had to be cut down with chain saws, piled up, and burned. That was quite a job. I can remember working six days a week with forty other men for an entire winter at that site, even when it was thirty below zero.

I worked eight hours, six days a week as a laborer. I worked very few Sundays. The contractors I worked for weren't on that tight of a schedule. Regardless, it was a domineering job. If people wanted to work harder and longer, they could. Some people were happy with forty hours, and others were not happy unless they got eighty. I couldn't work too many double shifts because my body got worn down and I was not productive. I was doing a lot of physical and manual work.

Ted Catanzarite

One of my jobs as a laborer on the project was assisting a welder. I gave him supplies and handed him rods. All of the concrete on the locks and dams was reinforced. It was poured in this very dense structure of reinforcing rods and he welded them together with cross bars. On the control tower on the main power dam there is almost as much steel as there is concrete. They had a lot of trouble with the steel when they got to the top. I guess it was an engineering nightmare. They were knocking this and banging that and changing this to try and get all the lines into the control room.

The winter of 1955–56 I solidified the bedrock at the power dam with liquid concrete that was pumped in with a pressurized hose. On the top of each machine was a meter and one worker had to stay there to make sure that the pressure was correct. He sat there with a wrench so that he could adjust it if it got too high. He opened it and squirted out some of the air to reduce the pressure and then closed it. I worked right through the winter. I had grown up there, so the cold didn't bother me. But it gets quite cold on the water. One time I was standing in mud because that was where the hole was that needed to be filled. I wasn't moving around a lot and it was about fifteen below zero in the middle of the day. My boots froze in the mud, but luckily not to my skin. I had to call for help. I got teased about that for a long time.

On Eisenhower Lock, I worked twelve hours a day. I worked for thirteen days straight and then had a few days off. As a young kid working long hours and not having a break didn't matter to me because I was making a lot of money. The minimum wage at that time was $1.05 an hour and I was getting $3.55. It helped me finance my college education.

Ron Barkley

When the painting contractor hired me, I painted the gates and cranes on the Long Sault Dam. If the carpenters or the millwrights were getting ready to set a generator at six o'clock the next morning, we had to paint it the night before. The supervisor would ask a few of us if we wanted to work a double shift. Most of the time two or three of us would stay on. I worked all kinds of hours; there was no forty-hour week. You worked just about all you wanted to work. The union rate when I started was $3.25 an hour. It was good pay for 1958 because factory workers were getting $0.87 an hour. The hours were long.

Tour guides and public safety officers held two of the most visible jobs on the project. Piquing the interest of and protecting the thousands of tourists who visited the construction sites each year affected the project's reputation. Guides conducted tours on both sides of the border, and the safety and enjoyment of those visitors meant they might come back or send their friends. The St. Lawrence Seaway and power dam construction project also stimulated interest worldwide. The project's impact on international trade and the defense of the United States intrigued dignitaries and engineers, especially those involved in rebuilding or improving their own country's navigational systems to compete in the increasingly global economy. National and international newspaper editors published daily photos of the project, updating Americans and people around the world on the construction progress. Robert Moses hired an official photographer to supply site pictures for PASNY's records and to distribute to various media outlets. As the largest construction project in the world from 1954 to 1958, the St. Lawrence Seaway and Power Project gained almost celebrity status.

Arthur Murphy

My primary job was to supervise Ontario Hydro's public relations' staff and conduct tours for special groups and dignitaries from Cornwall and from around the world. My previous experience working in the warehouse and cataloging pieces of equipment gave me a better insight into explaining the whole construction process. One of the things I showed visitors during the early stages of construction was the closing off of the Long Sault Rapids section, which for years was a rapidly moving section of the St. Lawrence River. Some flat-bottom boat pilots used to shoot the rapids. One of the interesting sidelights of the tour was the area that was going to be flooded on Inundation Day. You could see where construction workers were building beaches along the new lake and it would really marvel tourists because I would say that the water level would be up to that sand level eventually. You could see all the old valleys and the highways and all these interesting things that eventually would be covered. As the construction progressed, our tours were moved to the perimeter of the sites because we couldn't get too close to the heavy equipment. I still could point out the large cranes and how the dam was going to look when it was completed. After Inundation Day in 1958, I moved into the actual power house itself and was able to take people on tours of the bowels of the dam, including the control room.

I took special guests out to the middle of the power dam, which was an unmarked border between Canada and the U.S. Sometimes I called my counterpart at the Power Authority of the State of New York, and he would take me over to his side and show me his public relations department, his displays in the visitor's center, and how he was promoting his side of the power project. Then I took him back to my side and showed him our operations. Crossing the American and Canadian border was not a big deal. We knew our bounds and we had security guards.

Alfred Mellett

As the chief photographer for PASNY, I was in charge of all the publicity and photography for the project. I initially surveyed all of the power dam and related sites for Uhl, Hall, and Rich, the engineers hired by the Power Authority to be their on-site designers and managers. I had originally worked in Turkey for their parent company, Charles T. Main, laying out and supervising a power and tunnel project. When I came back the excavation on the St. Lawrence Project had started and I was sent up there to lay out all of the reference points for each dam site on the river. After two years, Robert Moses, the head of PASNY, hired me to photograph the progress of work on each of their sites and hand in weekly reports. I had the knowledge to write the reports because I had surveyed many of the project sites prior to construction. I was mainly interested in taking photos of the men at work. The subjects of my pictures were entirely up to me, so I set my own schedule based on when the major work was being completed at each site.

The six biggest American contractors all bid for work on the Seaway. In each weekly report I described the progress on the whole project from the Iroquois Control Dam to the Barnhart Island Power Dam, the Massena Intake, and all the channels in between that diverted the river. All of the contractors' supervisors knew me and shared all of their information with me. They never held anything back. We were really close. It was a fascinating job because so many problems came up every day that needed to be solved. The project was so unusual because nothing like it had ever been built before, so there was really no way the designers could plan all aspects of every site down to the last detail.

The resident engineer from Uhl, Hall, and Rich was really my boss. One day he got a little bit angry because I didn't get to work until 9:30 in the morning and the office opened at 8. I told him that I was out at the Iroquois Dam until midnight taking photographs. Robert Moses was a really good friend of mine and he stuck up for me in every way possible. He told my boss that I had his permission to determine my own work schedule because he wanted all phases of the project documented.

The job went on twenty-four hours a day, so I often photographed the construction at night with all the lights on. I also took aerial photographs from a Piper Cub. I could fly it myself, but I needed a pilot to get me in at the right angle that I wanted to photograph. I covered all the sites from the very beginning of the project. The most photographic was the Long Sault Dam because it had a curve in it. I would cover everything of interest and from various angles and at different times of day. The Seaway Authority, for instance, had a photographer covering their sections of the project, but he drove a stake in the ground and once every month took a series of photographs from the same spot. In contrast, my photographs covered the personnel, the people at work, and things of interest to the world. Once a month the PASNY publicity department selected my best pictures and compiled all of the information I had sent in and produced one report that they sent to the American and international media.

Floyd Grant
I patrolled the whole project and oversaw traffic control on town roads. Morrison-Knudsen hired several of us to protect their project sites and equipment from vandals and thieves. I worked one of the three eight-hour shifts, but I liked the night one best. On the other shifts, I fought traffic, especially in the morning when the shifts changed at seven at Alcoa and at the Seaway. There were three main intersections in town that were congested every day. At these locations either a stoplight was installed or one of our officers directed traffic.

All of the smaller construction companies had watchmen. So I checked in with them, especially on the night shift, to see if there was any trouble.

It was also crazy on the weekend with all of the tourists coming in to view the construction sites. A lot of interesting things happened. One Sunday a French couple approached my car because the woman had lost her pocketbook. While we were talking, a car drove up and a guy jumped out with her pocketbook in his hand. He had picked it up when he saw them drive away from one of the overlook sites without it. The traffic had held him up and he had finally caught up with them. There was over $100 Canadian in her wallet. I also had a lot of problems at night with people leaving the bars and driving drunk or being a nuisance on the sidewalk. I also investigated a few thefts. Sometimes young kids came out to the site in their cars and it was supposed to be off limits. It wasn't anything serious. The stealing was probably some of the workers. They would take stuff home from the construction company.

Overseeing the compensation of all of the contractors' employees on the Seaway and Power Project required the attention of numerous full-time payroll officers, their assistants, and supporting secretarial staffs. Each contractor maintained and collected time cards from each of his workers and turned them in on a weekly basis to the overseeing agency for payment. Then the respective payroll staff of these entities recorded all of the hours, determined the proper pay rate, and printed out and distributed the paychecks. The large amounts of overtime that many workers logged and the transient nature of the workforce and temporary reassignment of others to another work site made this a daunting task.

Hubert Miron

As the head payroll clerk for Ontario Hydro, I had two main responsibilities. First, every Thursday, I completed all the payroll sheets based on the information the contractors supplied on their time sheets. I then gave them to the women in the office who made out the checks. Secondly, on Friday, I traveled to all of the Canadian construction sites and distributed all of the checks to workers on the Canadian side of the project every Friday from Cornwall to Iroquois, including Hartshorne the house mover and his crew. I worked for three and a half years recording the payroll and distributing checks for Ontario Hydro. I worked a forty-hour week. At first I used to go out on the project every day. Then I had a time checker replace me and I stayed inside and posted all the time sheets. Each worker had a badge with a number on it to wear on Friday to get their pay. I started at seven in the morning and paid the men at the transformer station first. From there I went to Long Sault and even on to Sheik's Island. My last stop was at the Iroquois Lock and then I would be back at the office by four. Sometimes my boss would ask me to take one of the office staff out on my rounds to see the progress on the various sites.

After a while, I got to know the workers by their faces. But when the auditors came around, they had to have their badges on. The auditors from Toronto came to Cornwall twice a year and evaluated the payroll. I took them out on Fridays to show them how I delivered the paychecks. I remember going into one of the cemeteries they were relocating and there wasn't a soul there. The auditors asked me why we were there. I told them I had to pay the men. So I tooted the horn and all of these heads popped out of the graves. If they touched the bodies they got ten cents more an hour.

A lot of the men worked on weekends because they got double time. The supervisor had a hard time keeping track of some of their workers' hours because the bulldozers and equipment operators were all over the place. Some of them complained that they were not getting paid for all the hours they worked. I had a time checker who went to see the men who were complaining and tried to resolve it with their foreman. We thought some of them were trying to cheat. It was quite a job. I think I was one of the last cars to go on the cofferdam before they blew it up. I went to pay the men on Friday and the following week they blew it up to let the water in.

Workers faced dangerous working conditions daily. Many knew that one moment of bad judgment or carelessness could lead to their death. A friend's dying as the result of injuries from a fall became a reality. Many workers made the ultimate sacrifice to complete this project on time and on budget. While the number of fatalities on the St. Lawrence Seaway and Power Project was low compared to other hydroelectric and navigational endeavors, most of the accidents were gruesome and predominantly fatal. Because of the amount of large machinery, lack of safety equipment, long shifts, and inexperienced workers, many workers were ill-prepared to deal with the challenging conditions. The awareness of the experienced engineers and skilled workers who had previously completed large construction projects of the dangers of their occupations did not prevent them from completing their tasks on the Seaway and Power Project. Contractors anticipated a certain number of fatalities based on the project's overall cost. However, the severity of the accidents on the Seaway, based on the height of many of the structures, shocked even those workers who had witnessed accidents before. The long duration of the project also caused workers to form strong bonds and a reliance on each other for their safety that made the death of a comrade more difficult.

In contrast, reporters, government officials, and contractors promoted the St. Lawrence Seaway as the safest project of its kind from 1954 to 1958 based on the infrequent occurrence of accidents and fatalities. In April 1957, Deputy State Labor Commissioner Charles W. Halloran stated that "[t]he St. Lawrence River construction project was a model of achievement unique to the world's construction jobs because it has the lowest accident ratio of any job like it in the country."[2] According to an article in the Courier and Freeman, *careful attention to numerous safety precautions by both labor and management resulted in only three fatalities among four thousand workers and an accident frequency rate less than half the average for the heavy construction industry. The safety records established on the Eisenhower and Snell Locks and the Wiley-Dondero channel saved hundreds of thousands of dollars in wages and workmen's compensation and the elimination of lost time factors.[3] Sam Agati, the leader of the laborers' union, agreed that the job was as safe as the time called for. "The standards were different back then. We had meetings to instruct our members on where it was safe to work. Also, the job was inspected periodically by highly qualified New York State officials after 1955."[4] His union steward, Walter Gorrow, added, "There were very few accidents, considering the large amount of unskilled workers that were employed on the project."[5]*

Official Ontario Hydro reports and workers' descriptions paint a very different picture. A June 1958 monthly "Accident Statistical Report" for the St. Lawrence Power Project Labor Relations Association reported that ninety-nine accidents took place among nine contractors including Ontario Hydro. The injuries ranged from twenty-seven slips and falls to ten foreign objects in eye. In an article in the Chicago Daily News, *a reporter chronicled a sign posted at the Beauharnois Canal site that read "287 construction days, 84 accidents." He also reported that* on a project employing 15,700 workers, construction experts estimated that 3,000 would suffer burns, scrapes, broken bones, or heart attacks in the first year.[6]

Before the establishment of the Occupational Safety and Health Adminstra- tion (OSHA), contractors and lead agencies created their own safety programs and standards based on the lax state and federal standards. Ontario Hydro assem- bled a safety committee composed of the heads of all of their job sites to investigate all accidents and to determine the cause and how to prevent others from occurring in the future.

Lou Perini, the owner of Perini Construction, contractors on the Long Sault and Barnhart Island Dams and the Eisenhower and Snell Locks, implemented his own safety practices. His safety inspectors lectured new workers on how their safe work habits reflected their responsibilities to their families, fellow workers, and the financial success of Perini. The company also posted eight watchdogs at each of their work sites to prevent safety violations and unsafe work areas. They wanted to ensure that workers wore their hard hats, used properly secured ladders, and did not throw tools or unused materials from high places. The goal was to main- tain a work area free of debris and accidents.[7]

David Flewelling

While I worked on the Grasse River Lock, there was only one fatal acci- dent. A worker was electrocuted as the result of a frayed power cable. There was a mobile crane receiving power from that line that would tra- verse up and down at the bottom of the lock. The concrete crews and car- penters used that crane to pour concrete and move other materials. The leading cable to the crane got spliced and the exposed wires electrified some standing water. A poor man stepped in that particular puddle and the high voltage being conducted through it killed him instantly. Other workers hurt their backs or had minor injuries that healed over time, but most never missed a day of work. The workplace was full of hazards. I just had to be alert and pay attention at all times.

Bill Spriggs

At ten o'clock one night, a worker was adjusting one of the one-thousand- watt floodlights used to illuminate the lock work area and got electrocuted.

He fell to the ground and I personally worked on him for about forty minutes trying to regain his respiration. Then the doctor arrived. I was kneeling between this man's head and the doctor told me to turn him over. About forty people were standing around and watching all of this. The doctor reached into his shirt pocket, pulled out a scalpel, cut open the guy's chest, and started to massage his heart. Only four observers remained after that. I got home about midnight and thought about all of the men working around those big machines and all of the big rocks. Another fellow fell off the dam and landed fifty feet below. A trucker also drove off the dam. Many deaths occurred.

Corps inspectors had few safety measures to enforce back then. Most of the time, they made sure that masons and those doing high-level work wore safety belts and other workers had nothing in their lunch boxes. Taking risks was the way they got the job done. I had friends working in the trades and they would watch for me. Contractors paid people to watch for the Corps inspectors, so they could cut corners and they saved a lot of bucks.

Thomas Sherry

One guy I knew was electrocuted. He picked up a temporary electrical cord. It had a bare spot and he was standing on the wet ground. I was always standing in tons of water. There was a lot of temporary wiring where workers had attached lightbulbs to cords to illuminate certain work areas or to power equipment. The ironworkers and masons were always falling. I know one incident that happened on Grasse River in 1956. Maybe they kept it quiet. I don't know why. To say that on a job like that no one died is crazy. I have been on regular jobs and men fall and get injured all the time. The contractors employed thousands of carpenters, concrete workers, and laborers. When they trained me they made me wear a hard hat and steel-toed shoes. The contractors were safety conscious, but they could only do so much. You will have accidents no matter how hard you try.

One day I was sitting in a bosun chair filling bolt holes in the recently cured concrete and the crane operator raised the form with me underneath it. I hollered like hell and he stopped lifting it. That scared the hell out of me. Everybody was supposed to be away from the form when it was raised to the next level. But somehow no one paid attention to me. I was under there patching and all of a sudden the thing started moving and I was taking a ride up and I wasn't supposed to be going anywhere. I was supposed to be staying still. The crane operator stopped lifting the form so I could lower myself down and get the heck out of there. That was one of those screwups when there could have been an accident.

Ira Miller

The work was fast paced and the weather was often cold and damp. That made it difficult to keep a firm grip on some of the power tools I often used. One day I was loosening the bolts out of a concrete form the crew had poured earlier in the day with an impact wrench. The worker on the previous shift had been removing smaller bolts, so I had to change the chucks from a smaller one to a bigger one. The chucks were supposed to be attached to the wrench with a twenty-penny spike and hard-pinned. However, someone had loosened the connection and the chuck unexpectedly slipped off and hit the wrench trigger. This set the bit running in my hands. It spun in both of my hands as I tried to turn it off. The doctor put twenty-two stitches in my hand to close the wound, but I didn't miss any amount of time with that injury.

Another accident happened when I was working on the Long Sault Dam. There was an experienced local union millwright adjusting the gate guides when the final half of the structure was being constructed and water was flowing through the completed side. The contractor had to adjust those gates a few times during construction as the Corps altered its plans. The man slipped from the perch he was working from and went into the water. The contractor never even looked for his body. Since he was wearing brand new gardener overalls made of heavy duck material, he would have been weighed down in the water, making it hard for him to swim. They figured that the swift current of the water tore off all of his clothes and his limbs. There was nothing left of him. It was just like going into a great big mixer. He wouldn't have lasted long once he got in that cold water. I was working close by him and I didn't see him fall.

Frank Reynolds

Two men drowned on the project when I worked on it. One of them could have been a suicide. As soon as he saw someone go in the water, the job supervisor closed down the work on the site to find the body, which was only right. A lot of men also got into horrific car accidents. I went to work early one morning because I wanted to check the river elevation. There was a method I could use if I got there early enough when there wasn't too much refraction off the sun on the water. I told the guys to get in early so I could get a shot across the St. Lawrence River down by Iroquois Dam. That morning I was driving five miles outside of Ogdensburg at about 6:30. I came around a corner and there were bodies all over the road. There were three bodies on the pavement and two guys leaning up against a tree with blood running all over them. Dr. Depew, who was a surgeon at Hepburn Hospital, lived three miles from the crash site. He was out there taking care of all these people. It was a heck of an accident. Some of the guys were piled in one car going to work. There was another guy coming back from work in Massena and he was half loaded. They had a head-on collision. There were three accidents like that. A foreman also got crushed by a Euclid. He was guiding one truck driver to a dump location and another trucker didn't see him and backed right over him.

James Romano

The first accident I remember involved two guys who got swept through the big power dam. They were painting and they fell off the scaffolding and went right out through the gates. The water was so swift that they had no chance to save themselves. They were working without safety belts that would have held them onto the structure even if they had slipped. The contractors weren't too big on making us wear our safety equipment. They were just happy to find enough crazy guys to do this kind of dangerous work. Another night when I first started on the project, I was trucking gravel to the first cofferdam site. That was when I first realized how dangerous it was working on that project. The contractor had all of us truckers backing up right to the edge of the completed road and dumping our load into the water. I was the tenth trucker in line, and all of a sudden the current washed away all of the stone we had been dumping for several days. I got out of line and drove away. The contractor tried to save money by just dropping gravel into the water and avoiding building a more substantial and more expensive steel-cell cofferdam. The problem was the current flowed at twenty miles an hour. I moved on to another site because I didn't want my truck to end up in the water. After that I had a constant fear of being swept down the river. I also feared being trapped at the bottom of one of the lock sites and being drowned if one of the walls collapsed.

Joe Marmo

As the office manager in the Uhl, Hall, and Rich office, I received the weekly accident reports from the various contractors. My supervisors always told me that it was normal for men to get injured or killed on a project of this size. So I never thought it was abnormal. However, when contractors decide to have thousands of workers out there twenty-four hours a day, seven days a week in bad weather, things are liable to go wrong. Tom Lescovski, our safety engineer, took care of those things. He investigated all of the accidents and went out and talked to the site supervisors to determine how another such incident could be avoided. I did not get too involved. I was too busy and barely had enough time to do my own job. I do know that accidents killed twenty-five men on the project and I don't think anyone thought that was extraordinary. No one seemed to think it was surprising.

Ray Singleton

On the Long Sault Dam a bulldozer operator drove off the side of the structure into the water when they were converting the water around to complete the second half. The water was so swift, they didn't find his body until a week later. I think for the size of the jobs, they did pretty good. Accidents happened, don't get me wrong, but any job I worked on, for every dollar the designers anticipate so much of that going on. But for the amount of work we did, and the way we done it, I think the accident rate was reasonable.

Robert Carpenter

I remember one night I was up on the slope with Bill DeMorrow from Watertown, New York, and we looked down and a guy was lying in the mud below. I climbed down there and pulled him up by his hair. He was having an epileptic seizure and no one was helping him. His co-workers just ignored him and let him lie there. Of course he should have never been on the job. He didn't want anyone to know he had epilepsy or he would have lost his job. One morning another guy came in and backed right over the fill that had been removed by the excavating crew the night before and the tractor he was operating rolled over and killed him. The men that I worked with had no interest in fringe benefits. They were a completely different breed. They were willing to make the ultimate sacrifice to complete the world's largest waterway project.

Ron Barkley

The safety conditions in the painting business only started to get better in the 1960s. The ventilation and safety conditions on the Iroquois and Long Sault Dam sites when I was painting the doors and railings were atrocious. It was worse than working in the mines. What I had was not even fit to be called safety equipment. Nobody put pressure on me to wear the little bit of stuff that I did have. That is why a lot of people are dead today. I had a rag to wrap around my head and mouth. I worked with safety ropes and the bosun chair to keep me from falling. But the masks for painting and sandblasting were not the greatest equipment going.

A lot of accidents also happened on the project sites. One of the most common accidents was workers falling off the top of one of the hundred-foot locks or dams, and landing on the concrete below. Some of the men that fell jumped up after they landed and said, "God, how far did I fall?" and then ten minutes later fell over dead because they were all broken up inside. One carpenter fell off the Long Sault Dam and he jumped right up like he hadn't even been hurt. The whole back part of his head was bashed in and he was dead eight minutes later. But he was talking and was awake the whole time. He was one of the last ones to be killed. I heard of a few other guys who had been killed. One fell asleep during his lunch hour behind a Euclid and someone backed over him. The truckers left the

Euclids running all the time, so if he was leaning on it, he would have never known if it had started moving. The truck driver backed over him before he realized he was there.

David Manley

I don't think we had good safety equipment back then. We should have been wearing goggles or earmuffs. Everybody had hard hats. Before the explosion crews set off a blast, a whistle went off signaling all of us workers to take cover or get in a hole. One time, a bunch of us got behind a great big crane. It was crowded and there was a guy whose head was sticking out and a big stone came down and split his hardhat and all he got was

22. Eisenhower Lock under construction in May 1957. Euclids speeding along the adjacent roadway are visible on the right of the picture. The observation area parking lot is in the upper left corner and has already been paved. Courtesy of Alfred Mellett and PASNY.

just a couple of scratches. Another night during my shift a guy fell off the back wall on the Eisenhower Lock. He fell about thirty feet, but landed in the water so he didn't get hurt too bad. He just injured his back.

Safety people patrolled the sites, but if a man wanted to get up to a pour quickly, he found a shortcut. He would put one foot on a hook and a crane would take him. Others rode up to the forms they were working on in the empty cement buckets, dangling their feet over the edge. It was lucky that more workers did not get killed. Some of the things we did there were crazy and the work site was full of hazards. Carpenters and concrete crews left piles of wood and broken concrete forms on the ground. No one wore safety belts when they were climbing up the lock walls. We all thought we were macho men back then. I was seventeen and I didn't think I was going to die.

Ted Catanzarite

People died and got injured on that project. On the Eisenhower Lock, one man had a heart attack. They had to get him off the structure by using one of those huge buckets which they transported the concrete in. I remember them lowering him down and carrying him off-site in a stretcher to a waiting ambulance. I don't know if he had a preexisting problem or if it was work-related. Two other people died on the bridge that they built to bring supplies and machines to Barnhart Island during the construction of the main power dam. I don't know if they fell off the structure or died from an injury. I was really too young to personally worry about the dangers.

Arthur Murphy

Many men working on the Canadian side of the power dam met an untimely death. It wasn't revered and it was not like it was unexpected because the safety standards and the safety rules for construction at that particular time were seldom enforced. The last twenty-five years of my career, I was involved in construction safety for Ontario Hydro. You can't compare the rules in the 1950s to the precautions we were taking when I retired in 1993. Heck, men worked on the power dam and town sites without hard hats; they wore fedoras. Many masons did not wear eye protection while doing stone work and drilling. It wasn't a big issue. It was careless, but that is a fact.

Some sad fatalities happened during the construction of the project which are embedded in my mind forever. I remember one incident where a lad was working aloft and he lost his balance and fell about fifty feet. He became impaled on a rebar, a piece of round reinforcing steel that goes into concrete for reinforcement. He was extricated and put in the site ambulance with the rebar right through him. He lived for about ten hours, but all of his internal organs were damaged. It went through his bottom end and he was still conscious. It was tragic. That was one of the more serious ones I was involved with and it wasn't very pleasant.

Even though PASNY and Ontario Hydro tried to prevent accidents by defining standards and safety regulations, in order to get the site built there was so much pressure put on contractors and employees to move ahead as quickly as possible. Supervisors and their workers knew the rules, but sometimes getting from point A to point B was not all that easy and people were put in harm's way. Ontario Hydro officials and all of their contractors knew there were dangers, but somebody had to do it. Painters were also exposed to harmful fumes while using epoxy paint. Today painters are required to wear respirators because there are carcinogens included in a lot of those products. We didn't know that men were being poisoned by the products they used and if scientists were aware of the dangers, they weren't sharing those facts with the workforce. So a lot of men made sacrifices to build the power dam. The total number of fatalities will probably never be known. Men went beyond the call of duty to get the job done. Contractors successfully completed the Seaway and power dam, but some of their employees met their demise.

William Rutley

When I worked on the surveying job for Ontario Hydro on towns one and two, I never knew of any serious injuries, just a few minor ones. Hydro was a safe company. They had their own safety men. If they saw a truck driver or an equipment operator doing something he wasn't supposed to do or that wasn't kosher, they would haul his butt in and give him a couple of days off. Hydro's safety inspectors wouldn't put up with it. Sometimes, if you knew the supervisor, he would give you a warning for the first violation, but if it was the second time you were sent down the road. They just wouldn't take a chance with a person's life. Ontario Hydro administrators focused on safety right after the war because they had lost a lot of people on the Ottawa River project. Their inspectors cracked down and made sure that men operated equipment properly and put up sufficient scaffolds. They didn't want to kill any more workers than they could help.

When I worked for Atlas Constructor, two of their employees were killed. A flagman for a Marion shovel was down in a hole they were digging and he heard a noise. The brakes had failed on a DW21. It was a two-wheeled cart that pulled a dump truck behind it and was steered by friction in the back wheels. The supervisors figured that something happened to the braking system. It shouldn't have happened because the DW21 had air brakes and the operators had been parking them on an incline for a while when they were not in use with no problems. Somehow the brakes released. When it started to roll down the embankment, the flagman jumped underneath the Marion shovel to save himself but something hit him. At first, they didn't think the man was hurt that bad, but he died the next morning of a serious head injury. They brought him into the warehouse and I looked after him until the ambulance came. Another worker was flagging for a Euclid operator that was dumping soil. When the Euclid operator took a break, the flagman sat down on a rock and fell asleep and one of the other truckers ran over him. Even though the project was big, there were not a lot of deaths. For the amount of man-hours, it was safe.

Hubert Miron

I remember the men who worked for Hartshorne, the house movers, started a fire on really cold days. Ten men would work for a few hours, while the other ten stayed around the fire to keep warm. Then they switched around. One day the supervisor told a guy to go home and change into warmer clothes because he wasn't dressed properly. The next thing I knew, he threw himself on the fire. He took a spell. He was all right. His fellow workers picked him up and sent him to the hospital.

George Haineault

I witnessed two accidents during the three years I worked on the project. I had a friend, a field guy from the East Coast, who was a signal man for the crane operator. There were two gantries on the Canadian side of the dam on large tracks. There was eighty feet between each track. The cab was mounted on top. The crane had a 180-foot boom on it. My friend's job was to signal the crane operator to lower the cement bucket or if we were placing a form to nudge it a bit so we could align our sleeve. To get to the section of the dam we were working on, we climbed up these twenty-four-inch-wide and twelve-inch-deep steel rods. Masons welded the rods into the cement at a forty-five-degree angle every twelve inches between the top and the bottom of the gate to make sure the channel irons were properly aligned. Hydro did not mean for workers to use this as a makeshift ladder, but since it was there, we used it to get from the bottom of the dam to the top quickly. What happened was the saddest thing. My friend started going up this horizontal rebar quickly in order to unhitch a load that the crane operators were lifting into place at four in the morning. When he reached for the top rung, the weld gave way and he went straight down to the bottom of the dam with that rod in his hands. The weld must have been fractured or damaged when it was lying on the ground, before it was installed. I heard about the accident when I arrived for work in the morning. It just about killed me. It made me sick to my stomach. It was a guy I knew well and often talked with. Within forty-eight hours, the bosses ordered welders to fasten rods going vertically up the inverted tracks that were mounted on the structure to form a sturdy ladder. It was too late for the man who was killed, but Hydro was unaware of the danger until the accident happened.

A young guy from Cornwall whom I went to school [with] fell only twelve feet, but he landed on a pile of steel and broke his back. He was crippled for life. I knew I would see him around town in his wheelchair for the rest of my life. It was a perpetually rotten memory. He was such a nice guy. It always happens to the nice guys. The heights were so high, it was unreal.

I pretty nearly had my clock done in by another guy that I had gone to school with. In the winter, in order to maintain some heat, we covered the opening of our work area with a sheet of three-quarter-inch plywood. Hydro provided us with heavy jackets like the ones the utility workers wear today. When he opened up the top, he dropped the sheet and it fell and brushed my shoulder, which was covered by my winter coat. When he came down, he was as white as a piece of paper. He asked me if I was okay. I thought he was going to have a heart attack.

Hydro had a very good safety record. However, I am a believer that all accidents can be prevented. People get careless. They call them accidents, but sometimes inexperience and exhaustion play a large role. It was dangerous work. There is no question about it.

Cyril Dumond

A few accidents took place when I was working on the main power dam. The one I remember most vividly involved one of our pipe fitters. He was working on a steam line and he thought someone had turned the pressure off. When he started loosening the flange the air behind it caused it to blow off with such force that it cut his head right off. There was another guy that died when a big cement drain rolled on him. The crane operator was dropping it in position and he was in the hole guiding it into place. The hook on the boom line broke and the pipe fell. The worker had no chance of avoiding being crushed against the side wall. Another night after I had left, two crane operators picked up a cab to mount on one of the big gantry cranes. The weight was too great and caused the two cranes to buckle up and fall on their sides. There was a carpenter named Lalone who was killed. They found him several days later crushed under the debris. He happened to be near the cranes at just the wrong time. Not all accidents, however, resulted in a worker being killed. One guy was electrocuted, but he didn't die. He touched a live wire when he was standing in water. There were puddles everywhere. Others injured their backs in a fall or cut their legs on a form or exposed steel rods. There were a few accidents, but not many. I knew five guys who died on the job.

Les Cruikshank

Men always get killed on a project this size. During the ten years I was employed by Ontario Hydro I was quite close to several fatalities. I helped pick up some guy at two o'clock in the morning. A truck had run over him. He was going up a ramp with one of these big old rock trucks, one of the Euclids, and he must have missed a gear or it came out of gear and he started to roll backwards. The steering wheel spun and knocked him out the door and the wheels rolled right over the top of him. He was a young fellow, in his twenties, who was getting married in the spring. Our superintendent got run over by a Euclid one time and he died on the way to the hospital.

Six guys also died when a Bailey bridge collapsed. Workers erected these temporary bridges to transport men and equipment to construction sites offshore. Mr. Bailey designed them during the war. The Seaway contractors put up longer Bailey bridges around the job than they had ever tried before and six workers fell into the water when one piece collapsed. Then four other men died on another little pedestrian bridge that collapsed. Some of these stories get pretty good after fifty years. Those things happen. A crane upsetting was not that unusual. Most of the operators were experienced, but there may have been experienced operators who took chances. Shit, I am guilty of that.

Bart Whitten

A day laborer was killed and after that the supervisors were not allowed to hire men on a daily basis. The temporary workers were not as alert to the dangers around them as the full-time workers who had been around large equipment before. This man was crushed as a crane fell over. It was at night and we thought more workers were underneath. We were lucky that it only smashed up two cranes. It took two days to get the crane cleaned up and thankfully only one man was discovered. The crane was one of the first deaths of at least twenty on the Canadian side. While the cofferdam was being built, a boat got tipped over when it was hit by some floating ice. Two men ended up in the water and because they were wearing too many tools they struggled to stay afloat. We found one body, but not the other. It was one of those really dark, wintry nights.

I was in an accident myself on February 13, 1957. A wooden form slipped and I was caught and crushed for a while. I was on my way to inspect a concrete wall that had been poured on the north end of the power house. The wooden form, which weighed a ton, slipped off the dam wall and fell next to me. It bounced off the ground and flew up and landed on my head, cracking the top of my hard hat. I don't think I would be talking to you today if I hadn't been wearing my hard hat. I was saved by my hard hat and given a very colorful one from the hat manufacturer. I became a member of the turtle club that day. That is what they called workers who were saved by the protective shell of their hard hats. I think I was the second man to have an accident like that on the project.

Considering the number of people on the project, I would have thought that more men would have been killed. You could not make a mistake around machines. On the Canadian side, I was constantly under cranes after my accident. It took me a month to go back on the job, and even then, I was always looking overhead. The Ontario Hydro safety people constantly devised new safety plans and studied all accidents once a week. They would recreate them and try to find out whether it was caused by fatigue or inexperience or if it was truly just an accident. This group came up with innovative strategies to prevent further accidents on the Cornwall Dam and future projects.

Arnold Shane
One of Iroquois Constructor's employees fell off the main power dam and landed on the steel rods embedded in the concrete below. Two of the rods went right through him, but didn't hit any organs, so he survived. He went on sick leave for nine months and eventually returned as a mail runner. I broke my hand. Back then truck manufacturers installed handles on the steering wheel of their bigger vehicles, making them easier to turn. One day I was driving on a bumpy piece of road and the handle spun around and broke my hand. That injury left me unable to drive a truck, so I went back to school. To treat minor injuries like mine, Ontario Hydro built a temporary hospital and purchased an ambulance. Medical personnel came right to the scene of an accident to take care of injured workers. If it was a really serious injury, they would send you straight to the hospital in Toronto, but if it was minor, then they would treat you at the on-site Hydro hospital. When I think back, there were accidents, but not a rash of them. It was just what happened.

John Moss

Most of the accidents occurred when a worker fell from the dam. One rigger had been there a long time and was getting married on the weekend. On Tuesday he fell to his death. He was a nice fellow from Quebec. There were a lot of high things that people were not accustomed to and this created a lot of hazards. All kinds of things caught you off guard. If you happened to be off guard, it didn't pay. The machinery was big and the job sites were high.

Many workers were also injured on the job and sent to a hospital in Toronto. I often heard several weeks later that they had died. Sometimes it was someone I didn't know personally because there were hundreds of men on the project working at various sites for different contractors. Other times the name would sound familiar, but I couldn't put a face with it. It was kind of something that happened. You would hear about it on the change of shift.

Joseph Couture

An accident occurred when two shovels with a boom were trying to erect a crane. It was not easy equipment to operate and the men may not have been properly trained. One operator turned the boom too fast and it put too much weight on the second one and caused the crane to collapse. Another day, steam blinded some laborers cleaning off the concrete forms with high-pressure hoses and the boom came around and hit one of them and he was killed. I only remember three men who died. I was always looking out for myself and keeping an eye on those around me. More accidents would have happened if Ontario Hydro was not a very well-run and secure organization.

A couple of guys were injured under my supervision. One fell on reinforcing steel and hurt his leg. I sent him to the hospital and the doctor sent him back to work even though he was limping. This was at end of the project and I was starting to lay people off. I had the option to keep some of them on. I asked this guy how his leg was and he said it hurt. So I took him to the hospital and met with the doctor. His treatment would not have been covered if he was let go by the company. I sent him to the compensation hospital in Toronto where he had an operation on his leg and spent a year there recuperating. He stopped in to see me in Cornwall on his way home on the train to thank me. The other man who was hurt spent nine months in the hospital and received his full pay during that time. I took special care of them and both recovered fully. If I hadn't watched out for them, both would have spent the rest of their lives as cripples.

5

Construction Dilemmas

THE HARNESSING of the Long Sault Rapids and the dredging of a navigable waterway by contractors on the St. Lawrence Seaway and Power Project has been described by engineers and designers as a triumph of man and technology over nature. For centuries local residents admired the Long Sault Rapids for their natural beauty. The swift undercurrent of this section of the St. Lawrence River challenged boaters, but held the promise of cheap power in the minds of politicians and engineers. After a half-century of political debates, catastrophic world conflicts, and economic strife, technology, machinery, and binational approval finally caught up with imagination as the Seaway and power dam construction began. By 1954 the Corps had completed designs to tame the current of the river and construct complementary power production facilities and navigational channels. The site contracts were awarded to the lowest bidders and bulldozers began to clear the lock and dam sites. Cofferdams large enough to hold back the water had been developed as had concrete mixtures that could be poured in larger amounts and withstand temperature variations. Gantry cranes and draglines that had been tested on other projects by the TVA and by the Corps assisted contractors in transporting supplies to high locations and allowed operators to remove difficult and voluminous amounts of underwater soil and material.

Problems surfaced, however, even before the first contract was let and continue into the present day. In 1942 there was a severe earthquake whose epicenter was at Massena Center. It knocked down every masonry chimney and headstone in Massena. A fault was located under the original site for the Snell Lock, so the Corps designers moved the structure to ensure it sat entirely on one side of the fault. The Franklin Delano Roosevelt–Robert Saunders Power Dam, however, straddles this line. Also, an expanded program of drilling and core samples uncovered substantial amounts of marine clay and glacial till near both locks and in all designated dredging sites. The rapid current of the river harnessed to produce power also was unpredictable during construction, and rapid water level changes often caught river engineers by surprise. Today many inexperienced boat captains run aground on elevated shoals during dry summer months.

The Corps and their contractors faced two major dilemmas on the dredging and lock projects: horrendous subsurface conditions and inadequate concrete supplies. Many contractors did not comprehend the difficulty of excavating marine clay and glacial till, and low bid their aspect of the project. Contractors reported cost overages and demanded time contingencies when they needed to blast most of the till and find creative ways of disposing of the clay. The shortage of portland cement led the Corps concrete division to approve the usage of natural cement on the Eisenhower Lock, which resulted in honeycombing during construction and the discovery of structural problems following the first season of lock operations. The river had always held her secrets, and once contractors and workers uncovered them, she fought to the end to maintain her original path and strength.

Garry Moore

Cornwall residents and politicians were concerned at the time they decided to build the dam across that particular section of the river because of the 1944 earthquake. I was ready to go to grade two that year and the school had fallen down the night before. So I didn't have to go. The funny part of it was the tombstones in the graveyards in Cornwall turned south and all the ones in Massena turned north, so therefore the epicenter of the earthquake was right down the middle of the river. Everybody was concerned with the dam being built across this earthquake fault. There was a lot of grouting to prevent water from seeping underneath the dam. Contractors discovered a lot of fissures and underwater caves that nobody knew about when they drilled and started to grout.

Keith Henry

I moved to Cornwall in February 1956 with twenty-nine months left before July 1, 1958—Inundation Day. Everything was well under way, but there was a lot of work to be done. In the International Rapids section, which was my main concern, many of the contracts affected one another by changing the levels and the varying currents. These changing conditions also interfered with the continuing navigation. The forty-mile stretch meant that the work was very spread out and made communication between the various sites difficult. Because each of these contracts was overseen by one of the four agencies, they decided they needed someone to regulate the water flow, and that was me. The contractors—the two power and two seaway entities—told me what they were doing and I had to make sure that all of the other affected parties were ready for changes that might affect their work. I was responsible for keeping Gordon Mitchell informed of the progress of all of the jobs in the International Rapids section. I spent a lot of my time reading the twenty-five river-level gauges in the area. I also had to be aware of construction progress and the exact river flows. I paid attention to the failure of contractors to stick to the schedule and suggested how to compensate them for different conditions they encountered and how to get other contractors to agree with these changes. The overriding consideration was that everything had to be done by July 1, 1958, so that the flooding could take place. The critical date was always in our minds.

The first and worst seiche (an occasional change in the water level caused by wind, earthquakes, or change in barometric pressure) I encountered during construction was in June 1957. Some of the channel enlargements on the upper river had been completed, while another was just beginning to affect the river gradients. As a result the river levels in the International Rapids section were elevated above their normal range. On June 29 I was leaving for my annual vacation. On Saturday a big positive seiche occurred. My assistant had gone to Montreal for the day and when the accompanying storm hit the Montreal area a number of the street were flooded on the west end and he couldn't get back to Cornwall for twelve hours. I was busy moving supplies into our rented cottage on the Ottawa River and the only person Mitchell could get ahold of was Marg. She eventually got me a message that I needed to return to Cornwall immediately.

When I got back I found that the seiche had dumped a lot of water into the St. Lawrence and that water levels were high above the Iroquois Dam. We quickly opened the structure and let the water through and did the same at the Long Sault. Eventually all of the water passed through the remaining dams and into the Gulf of St. Lawrence. We sorted it out in two days, but a lot of damage had been done to the canal banks that needed to be repaired. This was a reminder to us all that we were dealing with a man-made system that did not allow the river to operate the way nature had intended it.

As construction commenced, men on the job sites dealt with the day-to-day challenges. The cold weather and muddy conditions caused numerous mechanical and hydraulic issues for both contractors and independent equipment and machine operators. In the winter, oil froze in the engines overnight, making the vehicles inoperable. In the winter of 1955, contractors instituted special cold-weather maintenance programs, including chipping mud from the track rollers between shifts. Peter Kiewit and Johnson and Johnson, the main contractors on the Iroquois Dam, constructed heated sheds to store their equipment. They also used ether spray around the air cleaner as a starting aid on Mondays. Truck drivers and equipment operators were supplied with heaters to place under the radiators of equipment that was not gasoline powered. On the Grasse River Lock excavation, Dutcher Construction Corporation used heated diesel and gasoline to loosen dried clay on their equipment. Other contractors covered equipment with canvas to prevent the engines from freezing. Operating engineers often left equipment running between shifts or overnight for fear it would not restart. Contractors ordered workers to park backhoes and bulldozers on boards to prevent their treads from freezing to the ground. The mechanics also changed the oil after every one hundred hours of operation in all bulldozers, cranes, and earthmovers.

The wind also wreaked havoc. With windchill of fifty degrees below zero, material solidified in truck beds and became difficult to remove. Drivers created innovative methods to prevent the material from freezing in transit and site workers used gravity and bulldozers to loosen loads at the locks and dams. The harsh soil destroyed most of the trucks and backhoes driven on the project or greatly shortened their useful life. This damage cost contractors and independent drivers time and money in repair and shipping costs if new parts needed to be purchased from out-of-town suppliers. These challenging circumstances wore out shovel blades and men.

23. Construction at the Long Sault Dam continued in the subzero weather of February 1958. Truck drivers continued to transport material to the main dam site via the man-made road. Courtesy of Alfred Mellett and PASNY.

Robert Carpenter

One of my memorable moments involved Peter Kiewit, who was one of the major contractors on the project. He was a character! Kiewit visited the Corps of Engineers field office when I first started working in Massena and asked me to drive him out to a spot centrally located between the main power dam and the locks. At one point he told me to stop the car. On this plot of land, Kiewit wanted his workers to build a specially designed repair shop for his equipment. He drew a crude layout of the dimensions of the structure and where he wanted doors to be, so the operators could drive the equipment in to be repaired and then drive it out the other side.

Well, when he came back six months later, he asked me to drive with him to the completed structure. His employees had put the shop where he told them to. However, they had only installed doors on one side of the building. When we arrived they were driving a piece of equipment in for repairs and then turning it around and driving it out the same doors when it was done. When the repair man pulled in the next piece of equipment, Kiewit told him to drive it out the other side when he was done. The driver protested because there was no door there. When the worker refused to do what Kiewit asked, he drove his car right through the wall and buried it. We all just stood and looked at him. He said, "When I tell you to put a door in over here, I want you to put a door in over here!" I will never forget that. He wanted to show his employees that he was still in charge even when he was not physically there every day. I got a kick out it. That is the type of men that ran all of the construction companies. They never complained about how difficult something was. They just found a way to deal with it. The St. Lawrence Seaway and Power Project should be a memorial or a monument to all the workers and engineers who were unfazed and over-came all the insurmountable problems Mother Nature presented.

Ray Singleton

The Canadians finished their side of the power dam ahead of us because they poured in the wintertime. On the American side, the Power Author-ity wasn't prepared for wintertime concrete placement, so the contractors worked on excavation instead. Contractors laid off a lot of carpenters and laborers in the wintertime, but as an equipment operator I worked year-round. I operated a shovel on the Canadian side one winter digging on the islands. They were building a cable way to bring the rock across to dry up the Long Sault Rapids. I was involved in the operation to reroute the water around the project sites. The temperature was anywhere between 28 to 38 degrees for seven or eight weeks. If I was in the shovel, all the heat from the engine kept me warm.

Working on the Seaway project was different in terms of the weather for me because I am a southerner. It was colder than other project locations and I was working outside even when it was below zero. Those of us who stayed just adapted. I had worked all over in places where it was as cold as you can get and just as hot as you could get. However, cold weather affects your equipment just as much as the extreme heat. One day it was fifty-three degrees below zero and all the equipment was turned off. It took us three days to get it started. When the temperature was below freezing, you couldn't lay your hand on anything metal because your fingers stuck right to it.

Equipment operators also couldn't leave a bulldozer or any piece of equipment with tracks on it in a muddy or watery hole because it would freeze to the ground. If I didn't clean the tracks out at night, I would go back the next morning and the tracks would be frozen in place. There is nothing any stronger than frost. I often left my equipment running for the operator on the next shift. I had someone walk around and watch it. I couldn't turn it off and let it get cold again. Contractors lost production time if operators were waiting for the mechanics to restart the equipment at the beginning of each shift. It was cheaper just to pay the cost of the fuel to keep them running continually.

The weather also caused the metal on the excavating shovels to become brittle. If the ground was frozen solid and they tried to excavate it, the shovel would break in half and the mechanics would have to weld it back together. Under these extreme conditions, the frost could be six feet deep. Contractors could not dig it or scrape it out with normal equipment, so they used drills and dynamite to loosen it.

James Romano

I was one of the few truckers who came for the winter season. During the summer I would work on the Thruway and come to Massena in the winter. There was a bunch of truckers from as far away as Long Island. Most of them would take off after the summer was over. That was when the contractors started to hire truckers. But it took a toll on your equipment, if you didn't know how to take care of it. Every morning ten to fifteen below zero was normal. The two coldest places in this country are Malone, New York, and Razor, Colorado. They are always in the cold storage. If I spit on the fender, the thing would turn to ice. You can tell it is cold when you walk and your feet squeak. I learned how to balance the oil and heat in the engine so it wouldn't freeze. I had to plug in a light and put it under the hood of all my trucks to keep a little heat inside the engine block. Other truckers had block heaters that they would plug in. I used to get up in the morning and check the oil. If my truck didn't start on the first try, I forgot about it.

I finally rented a gas station and kept two trucks in there at night. Then in morning when those two trucks started I pulled in two more to get them going. Then I had times when the oil was like cement in the motor. The oil was so heavy my motor would knock because it wasn't getting the oil it needed. We didn't have the winterized oil that we have today. It was a straight twenty, thirty, or forty weight. The equipment operators also learned how to deal with the cold. If they parked a piece of machinery on the ground, it would freeze in place. Then they would have to melt the ice with a gasoline-powered flame-throwing weed burner.

The material I was hauling often froze in the back of my truck because the windchill in the dump increased as I was driving. Going fifty miles an hour when it was twenty degrees below zero made the windchill factor one hundred degrees below zero. When I arrived at the site to dump my load, it was stuck in my truck and the counterweight would cause the truck to tip up. So I started spraying my truck bed with kerosene or calcium. Other guys used to seal up the bottom of their dump truck and put their exhaust pipe into the chamber to heat the bottom of the bed, so it would stay warm and the stuff would slide out. If you tried to dump your load and it was stuck, the guy would come with a grade hauler and pull it out of your truck. A couple of trucks just rolled right over. The drivers were lucky they didn't get killed. I had two pistons on my truck, so it was easier to balance.

Ron Barkley

I initially drove a truck on the Iroquois Dam project. Men from New York City started trucking companies by buying four or five trucks. They had heard that there was a lot of money to be made. The big problem was that they didn't know the first thing about running a business and they let anyone drive their trucks. They lost a lot of money and most of their trucks were destroyed because they did not hire any mechanics who could take care of the vehicles. I was hired by a guy who bought three trucks and sent them to Massena without accompanying them. I worked hauling dirt when they removed the cofferdams. My fellow drivers were in the ditch all the time tipped over because they were inexperienced. Some of them got into situations they couldn't get out of because they couldn't shift the trucks. The road was completely muddy and it rained almost every day. The clay that they were digging out of the river was slippery too. It was like driving kiddy cars in a carnival because you were going ten miles an hour at top speed and you were slipping and sliding and having fun. If I slid into the ditch, I had to wait two or three hours for the bulldozer to pull me out. It didn't matter how long it took because I was getting paid by the hour.

Sometimes the soil would not come out of the back of the trucks because it would form a vacuum and stick to the sides. I would lift up my tailgate and once it got to a certain point my truck stood right up and the front wheels would come right off the ground. I would be sitting there not able to do anything about it. Then all of a sudden that mud would let go and down came my truck. That happened to almost everybody. Of course the odd guy would forget to open his tailgate and then put the dump up and he would go up in the air. The supervisors would attach a cable onto a bulldozer and pull it down. The equipment was never any good after it left that job.

Eunice Barkley

When we were in the Syracuse area, my mother worked on the New York State Thruway with a lunch wagon. When the Seaway was starting she wanted to move back to Massena because that was where we were from originally. I never finished high school. I only got through tenth grade because I had to work on the lunch wagon and help my mom with the two younger children, who were three and six years old. Mom also had a motel restaurant just outside of Cortland and it wasn't doing that well. It was open twenty-four hours a day, seven days a week, and she was just exhausted. So she decided she would take the lunch wagon to one of the sites on the Seaway. She talked to the Power Authority and she finally got permission to put it at Long Sault Dam as long as she did not go inside the gate. We got there early in the morning and left about two in the afternoon. Breakfast was our biggest meal. Mom would make egg sandwiches and coffee. For lunch we would have chili, hamburgers, and ham and cheese sandwiches, which were our highest price item, and of course sodas. My mom was a worker, God love her.

The construction people were great guys. Even the engineers were nice. I never had problems. Once in a while I would get a smart one, but not very often. Most of them were just happy that we were there. They would come up every now and then through the gate. I don't know how their lunch hours and breaks worked, but we would have a group make their orders and they would finish their break and then a few more would come out. They joked with me and said, "How do you do all this stuff?" I replied, "If you drink my mom's coffee, you can do anything!" We always had a running conversation going with anyone who came every day. The only time they didn't come was if they were sick and didn't come to work.

In the winter, we worked a couple of times when it was forty-six below zero. My mom said we made a commitment to the men and we are going to be there no matter how cold or tired we were. Sodas were served in bottles back then. I would open a bottle of soda and pass it out the window to the men and watch it freeze in my hand as it went out the window. It was horrible. I wore insulated boots, socks, and clothes on the lunch wagon, but if I stood for more than a few minutes in the same spot my feet would stick to the floor from the frost forming because of the steam inside the lunch wagon. There was always an upside down icicle underneath the coffee pot. Every now and then I knocked it off.

Thomas Rink

The worst part of the whole project was the weather. It was a big factor all times of year. I can remember working when it was thirty below zero and the oilers putting kerosene in the equipment engines to make them work. When I worked on the Eisenhower Lock control towers one winter, we had to heat all the concrete. I had to keep everything warm for the masons along with another laborer. The first thing I did in the morning was to make sure I had the salamanders going and the scaffolding and the tarps up. Then I mixed the mortar for the masons and brought them their blocks. I had twelve masons and they were all different. Some of them wanted the mortar a bit wetter and some wanted it a bit drier. I also had to make sure to raise the scaffolds and the planks up for them to stand on. I worked my ass off. My co-worker and I mixed our own mud and cured it in pails and we took them up with pulleys and kept our work area warm because if any of the mortar froze, it wouldn't set. The contractors never wanted to go over budget. So if they could find two guys that could do the work of four, they were happy.

Neil McKenna
It was pretty bad working through the winter there. It was hard getting back and forth to work. The contractor I initially worked for didn't pour concrete in the wintertime. Instead we did a lot of office work, plotting cross sections and doing tidying-up work. I also did a lot of sitting in the carryall truck during the rest of the year. If it was raining, we couldn't do much. When I first started, I was laying out the east-west highway between the Eisenhower and Snell Locks. Somehow my crew and I got the wrong coordinates for the curve that left Eisenhower Lock going east. We went back and redid that part of it. Most of the time, it was a one-shot deal. We laid out a center line on a dike or a cross section on a certain part of the canal. The engineers provided us with drawings, directions, and coordinates which we needed to locate and mark.

David Manley
Getting drinking water was a problem, especially in the hot weather because the guys who made the rounds with the drinking water had three other sites to go to and their tanks would get empty fast. On the hot summer days, I would sweat a lot and I got really thirsty. Sometimes when the men who drove the water trucks didn't show up for work, we had nothing to drink at all. Then in April it was cold and windy working the midnight shift. I was right out in the open and high in the air on many of the lifts. It didn't bother me much. Back then nothing bothered me. After my shift, I went home, took a shower, and went to school. I would miss my first class, which was usually study hall, and then leave around 2:30 in the afternoon

and go home to sleep. Then I would get up, eat, and go back to work again at midnight.

Ira Miller

When I first started working on the Long Sault Dam, it was my job to drive an army duck hauling trash back from the dam site to the mainland. Attached to the boat with a half-inch steel chain was a trash boom made out of twelve-inch-square timbers that were bolted together with a twelve-inch spacer in between them. The whole thing could have gone over the dam pretty easy, but it had a lot of power. My boss and I were removing old sticks and logs that were caught in the dam. Sometimes we pulled big bolts out of the concrete with a steel chain and an axe. Other times, I would pull up to the side of the dam and we would clean up the trash and haul it back to shore.

I was used to working in the cold. I could take it then. The southerners did quite a lot of squawking about the weather. They should have known they weren't going to Florida. The winter of 1957 was probably the coldest in history. That wind that came down that old St. Lawrence River would just about scare you. I was right out in the open with nothing to break the wind.

Jack Bryant

There were some contingencies built into the mandatory completion commitments in each of the site contracts, but they were quickly used up. Thereafter, there was extraordinary pressure to meet the dates, no matter what the weather! Thus, where excusable delays were encountered—and there were many—the contractors were entitled to time extensions, but in most cases extensions were not allowed. In other words, notwithstanding legitimate extras, changes, unusually bad weather, and the like, which justified more time, the contractors were directed to meet the unchanged original completion dates. This amounted to acceleration of the work to fit in the originally allowed time, rather than the extended time to which the contractor was entitled. This, in turn, forced more and more work into extreme weather conditions when the contractors would otherwise be shut down. Winter work in the project area, particularly on a large scale, was poorly productive and hugely expensive. The contractors were ultimately paid for a large share of the added cost of winter work on two bases. First, they were entitled to be paid for the actual cost of added work and changes when done under winter conditions. Second, where excusable delays pushed original contract work into winter, and it was nevertheless required to be performed in winter rather than having a time extension until good conditions returned, the added costs of doing the work in winter were compensable. Some time extensions were eventually allowed but payment for extra materials and wages were not paid out until final contract settlements were arranged after the project's completion.

Garry Moore

Neither country really had a heck of a lot of experience in winter construction. Normally, contractors shut down their work sites from October to March, but the stringent schedule and complexities of the Seaway and Power Project forced them to keep going year-round. Most of them had to learn how to handle the cold weather and understand how it affected concrete and machinery. The experience level between the Canadians and the Americans coming into the project in terms of working in cold weather conditions definitely differed.

William Rutley

Cornwall is one of the mildest places you will find in Canada. I can remember in January 1955 it was fifty-five below and there weren't many vehicles that started that day. I had a little 1952 Chevy and I got into Cornwall, but there were not a lot of people at work that day. I saw a few of the operators blow the engine on a grader in Long Sault; they were trying to tow it to get it started. At that time there was no starter fluid. That was a cold morning and the wind was blowing too. It was a terrible day. I could hardly keep warm in the warehouse. I had an old coal furnace that Atlas put in so it would be comfortable for me to work. I sat near it and draped the rest of the warehouse off where the equipment was repaired. Even with that I had a hard time keeping warm.

There were lots of breakdowns when they started because the weather conditions and the twenty-four-hour work schedule were all new for some of the contractors. Many of them soon learned that if they operated a piece of equipment ten hours a day, six days a week, it wore out quickly.

Cyril Dumond

Oh my God, in the winter it was cold! When I started, a friend of mine, Jamie LaFrance, and I worked there all winter before we got a full crew. We were there on a temporary basis to provide whatever pipe-fitting work was needed. When we weren't needed, we climbed into a cement bucket and built a fire with scrap wood to keep warm. If they needed us for a small job, they would come and get us. Then in the spring, Ontario Hydro construction crews built boilers for temporary heat and lunch rooms. The first winter was very hard. It was cold, but we were well dressed.

Glenn Dafoe

We worked in every kind of weather. If it was raining, I was issued a raincoat and rubber boots. A lot of the homes that were moved needed new roofs and we shingled them in the wintertime. I tell people today, they don't realize how lucky they are. Today tradesmen stop working and are sent home once the temperature drops below a certain level. My employer just provided windbreakers and we never stopped. If I didn't like it, I could go home. I appreciated the job. Before I started working for Hydro I was getting $1.60 an hour. I started there at $2.20 and got time and a half for overtime.

The Corps dealt with their own unique dilemmas on the Seaway project based on incomplete plans and unanticipated soil removal and dredging issues. For over a century, the U.S. Army Corps of Engineers' administrative and construction staff had designed, built, and maintained most of the navigational projects in the Great Lakes. Their detailed blueprints and contracts, the reliability of their soil testing, and the quality of their completed locks and dredging projects were legendary. With the approval of the St. Lawrence Seaway and Power Project by the U.S. and Canadian governments, the Corps leaders, including the chief of engineers, Lieutenant General Samuel D. Sturgis, Jr., expected that it would be business as usual.

The Corps anticipated being the lead agency on both segments of the project based on their expertise and completed research. But the political environment shifted toward cost efficiency and the diversification of responsibility for the construction and maintenance of power and navigational routes. Based on that fact, the Corps administrators found themselves in an unfamiliar role on the Seaway and Power Project: advisor to two agencies—a provincial and a state government—to their contractors, and to the American Congress. Sturgis struggled to deal with the inexperienced administrators of the newly created St. Lawrence Seaway Development Corporation (SLSDC) and PASNY, which had never completed a large public works project and were obsessed with completing the waterway and power dam as quickly and cheaply as possible.

From 1954 to 1958, the Corps on-site inspectors and regional office managers maintained their stringent design and inspection procedures under an unforgiving time schedule, facing concrete, steel, and skilled worker shortages, and difficult subsurface and weather conditions. They did not want to be held accountable for an over-budget project hampered by missed deadlines. At the administrative level, Sturgis anticipated congressional opposition and critical public opinion if he asked for additional funding or time contingencies. In this tense and complicated environment, the Corps and their contractors strove to complete a well-built and highly visible navigational and power project. Initially the fact that the SLSDC decisions about cost and time frame impeded the proper design and completion of structurally sound dams and locks bothered Sturgis. To protect their reputation, the engineers from the Buffalo District Office of the Corps drafted three new design memos and commenced more extensive subsurface investigations in the summer of 1954. The Corps also outlined their inspection procedures and wanted the SLSDC to remain as a general supervisor and not interfere with field work. However, Sturgis lost a battle to retain the original St. Lawrence Seaway and Power Project completion date of spring 1959 because of the desire of Power Authority Chairman Robert Moses for the Robert Moses–Robert H. Saunders Power Dam to commence the generation of power by the summer of 1958 so he could begin repaying his bondholders.

Based on past experience, the Corps knew that there would be contractor contingencies if the typical work site changes, supply shortages, or strikes arose. However, Sturgis agreed after much debate to strive for an April 1958 Inundation Day with the understanding that based on unforeseen circumstances the date could be delayed until September 1958. This deadline fell eight months shy of the originally estimated completion date. However, even the best laid plans sometimes run afoul. Construction cost overages, contractor bankruptcies, and substandard materials plagued the contractors on the St. Lawrence Seaway and Power Project.[1]

Jack Bryant

Over the years, hundreds of soil and rock borings had been made in the project area by many entities including the United States Geology Service and the Corps of Engineers. PASNY also had a substantial drilling program. The subsurface studies were used to select construction locations, design structures, and estimate quantities of work. Boring logs and records of probes and test pits (when available) were included in bid documents and bidders used them to devise construction methods and estimate costs. The number, type, and extent of test borings varied, depending on the time and budget. The subsurface samplings cannot show everything, everywhere; to do so would require, in effect, doing the project before deciding what and where the project was to be. Sometimes borings from different sources and times did not agree.

The major problems on the St. Lawrence River project arose because there were never enough borings in the right places. Borings show the type, location, and quality of subsurface materials at the boring site. Between borings, you interpolate; beyond borings, you extrapolate. Both are expert guesses, but still guesses and not absolutes. The contract contemplates that the actual material encountered between borings will be similar. If it is not, and the actual material the excavation contractor encounters varies from what was indicated in the prebid information, the contractor can ask for a contingency or additional payment if the removal of the unanticipated material results in added cost or delay. In many cases during the channel dredging and lock excavation on the Seaway project, both occurred. This did not necessarily mean the subsurface data collected by Corps surveyors over the years was wrong; it meant that the data did not accurately predict all of the possible materials that contractors could anticipate encountering, and therefore they could not adequately estimate their costs in their bids.

Contractors bid for contracts with heavy reliance on the accuracy of contract data on soil conditions, site dimensions and required materials and add few of their own contingencies. Adding contingencies raises bids and loses contracts. Contractors are very familiar with equitable adjustment clauses which allow for payment of material and labor costs beyond a reasonable amount and rely on them. Generically, there were dozens of change orders negotiated for hundreds of items. There were a number of major changes in foundation areas where the subsurface rock was less stable and of different quality than anticipated. This required stopping work, investigations, added excavation (sometimes of the toothpick variety), added consolidation and curtain grouting, and negotiations to work out payment. Sometimes this required beefing up the design with heavier reinforcements. There were changes to incorporate better materials than thought adequate in the original designs.

Major contracts are based on detailed plans, specifications, and fine print that frame the contract obligations, both ways. Most large construction contracts have provisions for equitable adjustments in the contract price and time for four basic reasons: extra work, changes, changed conditions, and excusable delays. The PASNY contracts had such provisions. Federal contracts have similar provisions, as do those developed by the AIA (American Institute of Architects), NSPE (National Society of Professional Engineers), and others. The result is a balancing of risks and responsibilities between the contractor and the owner (PASNY). Extras are work or items not included in the original contract. Changes are alterations to the work originally included in the contract. Changed conditions are usually substantial variations in the subsurface or weather conditions from those contemplated in the contract. Excusable delays are those resulting from a number of enumerated causes, including extras, changes, changed conditions, strikes, embargos, unusually severe weather, and other causes not to be anticipated by a prudent bidder. Also included are excusable delays suffered by subcontractors and suppliers. A contractor encountering such delays is entitled to an equitable adjustment in the contract price and time for completion. Therein lay the monster of the St. Lawrence projects.

I remember an auditor named Shanley was employed by the Power Authority to audit everything that was done. He was perfect if you wanted someone to look out for your money. The man was absolutely convinced that if the engineers estimated that 870,000 yards of rock had to be removed and the contractor's final measurement was two cubic yards off, it was a mistake. That wasn't the case at all. The estimates were based on the data that was available at the time that the designs were put together and they were put together awfully fast. Then after each section of the power dam project was put out to contract to various parties, there were changes in the arc, and there were changes in subsurface conditions. There was no time to stop and think. We all just charged ahead. That was the Moses's way of doing things. It was very difficult.

One of the first jobs I got involved with was trying to get the kinks worked out with the contractor installing the power lines. The transmission towers were being installed in the winter, making the conditions virtually impossible. A. S. Wickstrum, the contractor, almost immediately threw up his hands and said, "This ain't the way I bid it. These are not the conditions that the plans showed." The Power Authority said, "We don't care, go ahead and build it!" I got in terrible fights with the job manager over the rates he was charging for the equipment. They brought in every piece of equipment they could find to try to cope with that mud including "aicky wagons," which had not been used much since the Boulder Dam was built. An aicky wagon is a trailer that is on crawler treads. They could drag those things around in the mud better than they could drag the wheeled equipment around. It was unbelievable. They wouldn't let them stop. The Power Authority officials told them to keep going.

The excavation contractor, Utah, on the power dam, went through every piece of equipment they owned. The marine clay burned out the transmissions of the big Euclid trucks that were hauling it off site because it was so heavy. The other stuff, the glacial till, ground down and wore out scrapers and shovel buckets. It was just worse than anyone expected and they were slow to learn. The contractors were also going through superintendents as well because the home office couldn't believe that they were producing that poorly. I remember Kiewit had a system a little different than the others. When they sent a bidding team to vie for one of the big projects, if they were successful the bidding team became the project engineer and administrative staff for at least the first year. That was a good idea because these men knew the most about the work they had contracted. Even the owner, Peter Kiewit, was really singing the blues at the Iroquois Dam site. I think it was one of the few contracts that was in his name only and the materials to be excavated, the availability of supplies, and the scheduling were not what he expected.

The marine clay and the glacial till fooled the contractors that bid. After the difficulties that Dravo had dredging the Long Sault Canal, many of the excavation contractors bidding for the Iroquois segment of the project, which included channel improvements as well as construction, came into my office to look at the test borings. The boxes held a conglomerate of material with lots of little rocks and stuff held together by clay. When they looked at the actual drilled pieces taken by the Corps during test borings, the material looked like ordinary rock. Based on these examples, it appeared to the contractors that they would be drilling through a gravel bed. When the excavation got under way and the material was completely different than the samples, they just could not accept how difficult it was to remove. Many had heard that Dravo was losing its shirt. Many sneered and said, "Well when we get in there, our operators will know how to do it the right way!" Well, when their bid was accepted and they got in there, they lost their asses too! It happened time after time. They ended up having to blast every bit of it because of the absolute tenacity of the stuff. I remember pictures of the glacial till excavation faces and it looked like just a big gravel bed. If you looked very carefully, you would notice that some of the individual rocks were a foot in diameter or more and were sheared through. That stuff in many ways was more difficult to get out than rock.

To excavate the marine clay, the operators tried an oversized, experimental bulldozer with a large blade on a long boom twenty feet out in front of it. They thought it would give them more traction as the tracks were on solid ground. But that didn't work very well either. Then excavation contractors brought in draglines including the "gentleman," which they walked from the West Virginia coal fields. Those pieces of machinery ended up being the only way that they could dredge the channels.

PASNY's and the Seaway contractors usually worked with specific plans and not through trial and error. Many would have preferred to take a few more years and do it right. On past projects, the Corps of Engineers designers found all the answers ahead of time and tried to adjust the timetable and plans for sites prior to putting the sites out for bid. The St. Lawrence Seaway and Power Dam Project was simply not built that way. Because of all of the entities involved, including Robert Moses, who managed to drive a lot of the contractors away, so many things went wrong. But the excavation contractors took the longest to understand the difficulty of the dredging. Some went bankrupt, while the wheeling and dealing of Moses and his old comrades from New York City pushed others to the limit.

Joe Marmo

The Corps had originally designed the construction of the whole Seaway and Power Project to be completed in ten years. They were going to run this project in a very orderly fashion, but I don't know if they would have done any better. If the engineers had a longer time frame, the designs would have been done and evaluated prior to the commencement of construction. When you do go into a project so unprepared, you risk the cost of change orders. We had tons and tons of changes and we had to negotiate each one of them with the contractors. In fact we got such a bad reputation, the contractors called us the "claims denial section." However, any contractor that had a lot of changes to the contract usually came out pretty good.

PASNY had the largest construction claim with the Perini Corporation from Framingham, Massachusetts. It was one of the first times I ever heard of this theory. It was called an impact claim. Perini wanted compensation for the impact that the design changes had on other elements. For example, if a concrete pour had to be redone and that affected all of the other pours surrounding it, they wanted payment to replace those as well if they did not align with the new pour. They convinced me of the merit of their argument and that they weren't full of it. It took almost a year and a half to settle the last claim and that kept me in Massena until the end of the project. It ended up PASNY settled the final claim with the Perini Corporation for $8 million.

Joseph Foley

If the Corps had been in charge and had more funding, I think there would have been fewer problems during construction. I could have hired more qualified inspectors, but this wasn't a Corps project. My colleagues designed the locks, outlined the various dredging projects, and supervised the actual construction for the Seaway Corporation. The Seaway Development Corporation accountants gave us a certain amount of money to complete each phase of our designated sites and Corps administrators could only spread that money around so much. We also had to share the construction duties and budget with other entities. The American Congress finally passed the Wiley and Dondero Act because planners set up a separate corporation to take the financial responsibility and long-term management of the facility away from the Corps. When you split the responsibility up among so many different organizations, it gets complicated.

All of the contractors and agencies involved in the St. Lawrence Seaway and Power Project had some real interesting problems. The soil conditions went from one extreme to the other. At least three contractors went bankrupt because the cost of removing these materials far exceeded their bids. I know the clay was pretty nasty stuff in a different way than the till. As long as the equipment operators dug out the marine clay when it was cold it was fine, but as soon as it got wet or warmed up, it was just like a big mud pie. It was difficult to remove, load into the dump trucks, and dispose of. In a way it was almost as bad as the till. The Corps knew about the existence of both of these substances. Our survey crews had taken borings up and down the river, but the contractors just didn't believe the till and clay were as hard as the borings showed. Our experts tried to tell them that they had done extensive testing and the results illustrated the tough excavating conditions. However, I don't think many of the Corps officials or the contractors believed that the soil conditions were as bad as the scientists made them out to be. I am not sure if the Corps had encountered these substances on other projects before, but it surely was not to this extent. All of the entities, including PASNY and the Corps, spent a good deal of their budget on dredging.

Jim Cotter
The Corps of Engineers didn't perform extensive core drillings. To conduct extensive subsurface investigations would have taken years. I don't think even in the bidding process the contractors had time to do any of their own testing. They just based their costs on the available information and the limited geological surveys. One of the greatest problems they had concerned the marine clay freezing in the truck beds in the wintertime. So the truckers rigged up a steam system. Each truck driver would stop at a station and pump steam through the actual framework of the truck so the load would slide out. I doubt if very many of them anticipated that many problems in advance. Issues arose every day.

Kenneth Hallock

The designers of all major construction projects run into significant changes because there is no way they can foresee all of the difficulties contractors may encounter, particularly underground conditions. The original excavation contractor, Jim and Jack Maser, on the Robinson Bay Lock, which is now called the Eisenhower Lock, went broke. I largely think that was because of the till. That particular material had not been encountered before. Contractors knew that it would be tough digging, but it was a matter of degree, particularly when they started the Long Sault Canal. Until they actually dug it up, they really didn't know the consistency or depth of the stuff. That turned out to be significantly tougher than anybody had anticipated and that resulted in dozens of contractor claims. Maser went broke and another contractor took over the job. It was the second contractor that came in and finished the digging. The excavation cost for that lock was higher than anticipated because it was tough work. It wore the equipment right out. In fact one of the sore points was that the excavation contractors ended up drilling and blasting, which they never anticipated. It cost more to blast and it was a whole different operation. But the glacial till was like solid rock in its natural state. The till also varied in size from large boulders to cobblestones. The excavators had to handle pretty much the two extremes. The glacial till was so hard and the marine clay was like toothpaste. The subsurface conditions were severe.

Roy Simonds

Excavation contractors dealt with difficult material including glacial till and green clay. The owners of Maser Brothers, the Badgett Mining Company, and later Tecon didn't realize the glacial till had to be blasted, so they clearly underbid that item. Many of them knew the hard rock existed because the Corps had done test borings and soil studies and made the results available for review during the bidding phase. Each contractors' specialists had the opportunity to go out and test the dirt and riverbed themselves at the proposed excavation and dredging sites.

The marine clay presented contractors with other issues. When the equipment operators excavated it and put it in trucks, it would stand up like asphalt. As soon as the driver started chugging down the road and the load started to shake, the clay turned into soup. Eventually the truckers drove the material to a spoils area where they dumped it out to dry. The fact that the marine clay was really packed in hard made it difficult to shovel. It ate through the equipment. The bigger contractors had dealt with projects of this size before, but had not run into those particular materials because they were unique to the St. Lawrence River Valley. I am sure the contractors became wary of it afterwards.

Ambrose Andre

On the Eisenhower Lock, the excavation contract was awarded before the Corps knew what they were going to build. The dug-out area ended up oversized. It was intended to err on the side of having enough room. When the excavation was started the exact dimensions and the thickness of the walls were not known. It turned out later that we had excavated an area bigger than what we needed. The contractors were told when to start digging a hole and when to stop.

Many of the excavation contractors, like Dutcher and Jack and Jim Maser, went bankrupt. Jim and Jack Maser notified the Corps in October 1955 that they were defaulting on their excavation contract on the Robinson Bay Lock site. Originally, Masers' bid was $300,000 below the government estimate and $100,000 lower than his nearest competitor's bid. What the Masers did not realize was that the estimated excavation cost was based on the usage of shovels and Euclids. They chose to use scrapers, which were a much more time-consuming and expensive way of removing the very difficult till and clay. The Masers quit after completing only half of the excavation work and Tecon took over. Tecon was owned by Clint Murchison, Jr., the son of a wealthy Texas oil tycoon. He also lost money. One of the owner's sons was in charge of the Tecon contracts. He had a thick Texas drawl. I don't know if he was learning how to run a job or whether he had done something wrong and this was his punishment. He was a misfit anyway. Nice guy, but didn't know much about excavating. It was a comedy of errors.

Both of these contractors did not know how to deal with the unique soils, so they encountered mechanical and financial difficulties. The glacial till was hard and almost like cement. It weighed as much as concrete. In it natural state, it was very, very hard. After it got wet, it turned to slippery grease. As most of Tecon's operators came from Texas, they had limited knowledge about how to deal with the northern mud. Many got their trucks stuck and had to hook a bulldozer onto it and try to drag it out. Then when that got stuck they brought in another bulldozer, which also got struck. Sometimes it took three bulldozers to move a truck. They also used a backhoe to dig the sticky mud out of the backs of the trucks.

Robert Carpenter

As far as challenges for the contractors, this was the hardest job I ever worked on because of the uniqueness of the underground conditions and the stringent timeline. It was a whole different ball game under pressure. The inexperience of many of the contractors showed in terms of their bidding and attempts to fulfill their contractual obligations on a project of this caliber. I don't think they knew what they were getting into. Maser Brothers from Pennsylvania, for example, was very green. They were trying to break into large public works construction. The company's owner knew the Seaway had a lot of different work sites up for bid and he wanted to be part of it. The Maser Brothers bid fifty-eight cents a square yard to excavate the Eisenhower Lock site. They shipped a whole fleet of modern scrapers up to Massena, which workers lowered down and pushed from behind with a bulldozer until they were full. Then the truckers came in and hauled off the soil or the operators opened the back of the scraper and dumped it in a designated area on site. However, the scrapers broke down and the mechanics couldn't keep them running. The operators also rigged up a big hook to a B-9 bulldozer that they would tow behind and break the dirt up with before the shovels came in. One of those hooks would last three hours before the maintenance crew arrived to sharpen or replace it.

Engineers also spent a good deal of time discussing the solid clay immediately downstream from Eisenhower Lock. Excavation contractors built a spoil area that took over a lot of property and huge pipelines from the dredges pumped the clay back there. The marine clay looked like chocolate pudding. Water was trapped in the membranes and when you picked up a handful and squeezed it, it turned to water. In other words, you couldn't stand on it or use it. But contractors built some of the dikes and the important structures on top of it where it was one hundred or more feet deep.

George Olsen, the Corps soil expert, and I performed a test to determine the shear strength of the clay and what we could use it for. We ran a vertical face right in it with a dragline until it failed. We had all these test holes behind it so we could measure the slippage. I got paid to work for forty-eight hours straight without a break to stay with it night and day until it failed because nobody knew when it would happen. Ultimately the contractors built a fifteen-foot-high dike to hold the water basin with the marine clay.

Operating engineers also removed millions of cubic yards of till. The Corps put a one-foot-square block of till from the proposed Eisenhower Lock site on display where contractors could look at it and do what they wanted with it. The laborer who cut that one-foot cube worked for two days, and went through two shovels and two pick axes. That was how hard it was. Reuben Haynes, one of the Corps soil experts, knew the tenacity of glacial till. I am not convinced the contractors had ever seen it or excavated it before. Over two thousand feet of ice had compressed it. When the excavators got down to bedrock, they discovered two lines across it where the glaciers had gone over it in two different directions. As a comparison, a cubic foot of concrete weighs 150 pounds a cubic foot, while the glacial till ranged from 165 to 175 pounds. It was more dense than concrete! In the end the contractors wound up drilling holes with rigging machines, packing them with explosives, and blasting every yard of it.

David Manley

The blue clay frustrated a lot of the contractors because the trucks kept coming to the dumping sites half full. The excavation crews filled the truck beds with wet clay, but when the workers opened the bed at the spoils area, the material had dried up and shrunk during transport. Some of the excavation contractors hired men called "bull hosers" to wash out the back of the trucks and get the clay out. When I got that stuff on my boots, it hardened and was difficult to get off. But in the work area where it was wet all the time, it was slippery and made the ground like walking on grease. The truckers got their tires stuck it in and then the blades of the equipment would fill up with blue clay. I don't think the excavators had as many problems with the rock because they could blast a lot of that.

Contractors borrowed stiff-leg derricks from companies who had worked on the Panama Canal. They also brought pieces of equipment to the Massena sites from coal mines in the West, like the "junior" and the "gentleman." They tried to use equipment from other canal projects, but most of it had to be adapted or replaced because it couldn't handle the glacial till and the marine clay.

Some of the contractors had never done this type of job before even though they constructed power dams and excavated channels around the world. The weather and the soil conditions were unique to the St. Lawrence River and the surrounding area. However, the contractors should have anticipated the soil and rock conditions because the Corps soils department had taken samples and displayed pieces of rock and containers of clay in their offices for prospective bidders to look at.

Garry Moore

Much like Corps contractors during the dredging from Cornwall, Ontario, to St. Aniset, Hydro equipment operators ran into sand, glacial till, and clay. Off Family Island rock had to be drilled and blasted. We ran into the whole spectrum. Our truck drivers also had to find places to put the stuff once we dug it up. Suction dredge workers removed the sand and the gooey stuff. Then our dredging supervisors cruised around in thirty-foot-long steel hull boats with echo sounders to find holes in the river they could fill. A lot of the stuff that came out by dipper dredge or ladder dredge went into dump scows and was towed to these areas that we buoyed off and filled within twenty feet of the high water level. Off Squaw Island near Lancaster where the sand area was, we created an island, but it disappeared the next year. It got washed away and went under water.

I remember the rock we ran into certainly didn't have any faults in it. It was pretty much good old solid Canadian granite. MacNamara, the excavation company on the Canadian section of the dredging near Cornwall, had drill scows. Each drill scow had two drills on it and a railroad track along one side of it. The two drills would move along the railroad track. The operators would start at either side of the river and move toward the middle. They would each drill about four holes on the bed, load them with nitric oxide, sound all kinds of whistles, and blow it. Then all the loosened material had to be taken out with clam buckets. Our boat operators patrolled upstream and downstream because the blasting stunned and killed whole bunches of fish. Many people thought this was a much easier way to catch fish than with a hook and sinker. Fishermen came out in their boats and lined up with their nets waiting to pick up the stunned fish. They were allowed to take the fish, but we had to keep them far enough away that they wouldn't get blown up themselves.

Keith Henry

The Corps had trouble finding excavation contractors. The contractors that were willing to take on a project of this magnitude with the severe soil conditions turned in higher bids than were expected and often over the government estimate. It was nobody's fault in particular, it was just that the contractors found out as each subsequent site was bid that the excavators on other sites were having difficulties and the removal of the material was not going to be easy.

Glacial till varies in different parts of the world and the great argument always concerns whether it is cemented or glacial. The American contractors claimed that it was cemented and that they had to drill and blast it like rock. It was very easy to drill and very easy to blast, but it cost money to do it. The Canadian contractors had very little trouble because they had been working up and down the river and they turned in higher bids for the excavation costs even on the first site. The American contractors came over to seek advice about the most successful excavation techniques and appropriate removal prices from the Canadian contractors. Basically it was harder than hell and they had to drill to loosen a lot of the stuff up before it could be removed. Digging it with shovels wasn't possible. Initially the Canadians had the same problems, but they had accepted that it was a difficult prospect early on and raised their bids accordingly. The Corps held their contractors to their original prices and paid them the difference in claims court after the project was completed.

As construction of the Seaway and Power Project commenced less than a year after a truce was signed for the Korean War, many construction materials including concrete were in short supply. Steel had been used to produce tanks and other

equipment for the war and other manufacturers had not converted their oper-
ations back to their nonwar products. The shortage of portland cement led the
Corps concrete division to approve the use of natural cement on the Eisenhower
Lock and several local concrete aggregate suppliers. Many contractors instructed
their workers to wear appropriate clothing to adapt to the varying climate, but did
not anticipate the impact these conditions had on concrete. Originally the Corps
plans did not allow the pouring of concrete in the winter. However, when the
construction schedule had been shortened by eight months, they amended their
policy. While temperatures soared into the nineties in the summer, the thermom-
eter plummeted to well below zero in the winter, causing concrete to crack.

After the first season of lock operations, Seaway officials discovered that the
concrete honeycombing, which they thought had been an isolated incident on the
Eisenhower Lock during construction, surfaced on the Snell Lock along with other
structural problems. Honeycombing occurs when air pockets form in drying concrete
based on improper vibrating and weaken the material's composition. In February
1985, the U.S. Army Corps of Engineers Buffalo District produced a "Reconnais-
sance Report for the Major Rehabilitation of the Eisenhower and Snell Locks for the
St. Lawrence Seaway Development Corporation." As of that date, contractors under
the supervision of the Corps had replaced sixteen of thirty-eight lock wall monoliths
on the Eisenhower Lock, installed post tension anchors to stabilize the walls on both
locks, and repaired cracked culverts at both sites at a cost of $15 billion. The remain-
ing walls on the Eisenhower Lock needed to be refaced and other areas on both locks
repaired. Without these alterations, the Corps estimated that one or both locks might
fail, causing the closure of the facility for one shipping season and a possible replace-
ment cost of $250 million. Given the many difficulties workers and engineers faced,
it is a miracle that the system still functions fifty years after its completion.

Neil McKenna

The winter I was there, the contractor had a major problem with the concrete
on the upper sill on the Eisenhower Lock. The pour over the vehicle tunnel
was full of honeycombs. I correctly placed the blasting caps and supervised
a safe explosion in the tunnel to break up the concrete, so my co-workers
could crawl in there and easily remove the defective material. The Corps
inspectors had informed my employer that due to the concrete imperfections, he had to blast off that part of the upper sill and then repour it at his

24. The concrete batch plant at Eisenhower Lock mixed all of the concrete on-site for the various contractors. Each mixture was specially ordered based on the size of the pour and the weather conditions. The final product was loaded into buckets that were transported to the lock site by trucks. Courtesy of David Flewelling.

own expense. As it was right in the middle of winter, we not only had to battle the cold weather, but make sure that our concrete mixture was frost-resistant and would not fail inspection the second time around. The limited space in the tunnel and the fact that it was full of an intricate maze of steel mesh and conduits made it a difficult work area. Concrete workers struggled to bring in the concrete and place it before the mixture hardened. Contractors did not pour a lot of concrete the winter I was there and that was the only part of any of the locks or dams that had to be redone. It was, however, a pretty expensive redo based on both the removal and rebuilding costs.

Roy Simonds
Perini and Iroquois Constructors both encountered concrete and foundation challenges on the main power dam. The Canadians built the project in a completely different way. They would pour concrete all winter long and the Americans poured twenty-four hours a day during the warm weather to catch up, but stopped in October. Iroquois also poured larger

lifts than we did, but it all matched up in the end. Both contractors also had to deal with the issue of grouting the bedrock under the dam. This was one of many surprises and extra costs which was not included in the initial contract or bid. All of these changes occurred because the Power Authority promised to finish the dam by September 1958 and went into the project with incomplete designs and test borings. Robert Moses believed that Uhl, Hall, and Rich supervisors could deal with the problems as they arose.

David Manley

Contractors on both the Eisenhower Lock and the main power dam encountered concrete problems. On-site concrete plant operators put the percentage of water, aggregate, and sand in the mixture and concrete crews did not vibrate the concrete correctly. One of the things I learned as part of a concrete crew was that you had to work the vibrators pretty good. If you didn't work the concrete well, it would honeycomb and the Corps concrete inspectors didn't like that.

Thomas Sherry

I remember my employer bought a lot of concrete from the Canadians. On top of that there was a lot of defective concrete installed on the Eisenhower Lock. After the Seaway opened the Corps had a hell of a lot of repairing because the faces of the monoliths on the locks were peeling and scaling off. The media never mentioned whether it was due to the mixture not being strong enough or whether it was poor design. In the winter the contractors covered both locks and emptied out the water. Friends of mine went down in there and patched the bad concrete. The Corps and their contractors would have guaranteed all of their work, so one of them had to foot the repair bill. It is very possible that someone was making some money by weakening the concrete mixture during construction. There was a lot of stuff like that going on. Someone was getting a kickback by filling the concrete with sand and greasing someone's palms.

During construction, the concrete batch plant workers must have done something wrong with the mix because concrete gets harder under the water if it is done right. Unless it is out in the ocean with something slamming against it, a structure should stay solid. The laborer would mix my concrete in a bucket which I used to fill the bolt holes. That stuff was very rich and very strong. On the larger pours, two men on the concrete crews used a big vibrator that looked like a torpedo to settle and work the aggregate. After the concrete dried and they stripped off the forms, all of the pours I saw looked beautiful, nice and smooth. But it took the test of time to find out how good it really was.

Kenneth Hallock

Soon after the concrete crew stripped off the forms on the concrete over the tunnel on the Eisenhower Lock, inspectors determined that it was faulty. So the Corps concrete division chief from Buffalo came up and looked at it with me to decide on the remedy. Based on the large amount of what we call honeycomb or rock pockets, we agreed that it needed to be redone. In other words, the matrix of concrete didn't fully fill the voids in between the pieces of stone aggregate. So we had the contractor take it out and replace it at his own expense. That was a big cost because there was six hundred cubic yards in that one lift. The contractor just acknowledged that the pour was too thick and full of honeycomb and it was his responsibility. So he had his men take it out and replace it. The general concrete deterioration only surfaced years later.

At the time the lock construction began, there was a major cement strike in the United States. Portland cement was in short supply. So we put into the specifications for both locks that the contractor could opt to substitute natural cement for up to one-quarter of the required cement content. Morrison-Knudsen elected to take that option, whereas the Grasse River contractors elected not to. They took a chance on getting concrete and they were able to do it. Eventually the Corps allowed the importation of Canadian concrete to be used on the locks. On the Eisenhower Lock, Morrison-Knudsen used natural cement and as it turned out the site concrete plants did not subject the natural cement to the same rigorous testing as the portland cement, whose exact mixture had been set by Ontario Hydro and Corps concrete labs. The practical effect of this was that the Eisenhower Lock concrete had only about three-quarters of the cement that it was intended to have. In other words, one-quarter of it was ineffective. It didn't cure properly, which weakened the concrete and made it susceptible to freeze/thaw damage. Through detailed examination of the concrete pours, we found ice crystal imprints within the concrete so we knew that the concrete had failed because it didn't have the strength it needed to resist freezing. Concrete has a tendency to gain strength rapidly at first and then gradually over a long period. That first winter some of the concrete was only a month or two old, so it had not yet attained its maximum strength and was susceptible to damage. This penetration took place within the first year, the winter of 1957–1958, because the concrete was left uncovered during that exceedingly cold winter. The Corps didn't permit natural cement to be used on certain sections of the locks for various reasons, and those areas where that restriction existed remain undamaged. So based on that fact, our concrete experts ascertained that the genus of this problem was the natural cement.

Designers of the Eisenhower Lock also didn't anticipate the effect of cold temperatures on the steel gates. The contractor installed them in the fall and left them exposed to the elements during the winter. Any material shrinks when it gets cold and those gates actually shrank in height about half an inch. The top hinges broke and I had to design a more flexible top hinge that could accommodate shrinkage.

Robert Carpenter

The Corps dewatered the locks every year to make repairs. One winter Ken Hallock called me in Atlanta and wanted to know if I had known of any difficulty with the cement content on the Eisenhower Lock during construction. I knew the contractors had poured concrete on shifts when it was forty below zero, but since concrete gives off its own heat, I didn't think it would hurt its integrity. Once the concrete workers completed a five-foot pour, they covered it with four-foot bales of hay to keep the heat in. But the forms on the side couldn't be covered and that may have caused some scaling right on the surface.

On the Eisenhower Tunnel the Corps concrete experts drilled into it afterwards to see if there was any honeycomb and there was quite a lot. That pour under the upper miter gate sill was the only one that was not properly vibrated. It took the contractor a month and a half to remove that pour and redo it, but it came out okay the next time. After that, the Corps changed the maximum size of the aggregate allowed in the mixture for various pours. For each section of the dams and the locks, the cement content, the air content, and the additive for workability were automatically cranked into each machine based on the size of the pour and the air temperature. There were three mixers at the batch plant, so every minute one of them was dumping into one of the trucks that would haul it into the locks. We had different standards and classifications than the Canadians, quick set, or sole set, for density. I just had to take it for granted that the people who knew cement and batching procedure had thoroughly checked it out. The scaling I think was the biggest item because it showed. Everybody would come and look at it and say, "What is wrong with Eisenhower Lock?" It was nothing dangerous because there was forty feet of concrete behind it.

Bill Spriggs

In 1959 I became director of the operation of the Eisenhower Lock. I had all of these maintenance issues. The concrete problems on the lock appeared during the first season of operation. Every year I held that position, the Corps dug out the pitted concrete and replaced it. It was bad. The Corps allowed the original contractor to use inferior concrete on the Eisenhower Lock. The Snell Lock was nothing like that because they used a different type of material on the walls.

The first shipping season in 1959 was a short one. The representatives from the Corps of Engineers who designed the structure and the contractors who built it were nowhere to be found and I had huge problems. The doors began to stick and the concrete started to crumple. In the winter of 1960, the Corps tarped off the lock and worked on the doors and new faces for the bearings. It has been a continuing struggle keeping both the Snell and the Eisenhower Lock operational. Each winter the Corps' contractors dug out a lot of concrete and replaced it. As an employee of the SLSDC, I was left to operate a flawed facility.

Ambrose Andre

In 1961 they closed the gates on the Eisenhower Lock because the water kept leaking out. They dropped the pool to a lower level and sent a scuba diver down and he said he could put his hand under the lower sill and it had raised up about six inches and floated downstream about a foot. It leaked like a sieve. So we delayed the opening of navigation for about ten days to grout the crack and reinstall the door. Then the following winter the whole sill was removed and replaced. One time when they wanted to change the bearings on the gates one of my buddies who was the chief engineer for the Seaway for years called me up and wanted to know how they could do it. It turned out if you put head behind the gate, it would stand there without being attached to anything. It took me a while to convince him of that fact.

Joseph Foley

The part that always bothered me more than anything else was the Corps designers and construction crews always did everything by the book and according to the cement specifications. The Seaway locks turned out with so many mistakes. I know we had several lists of concrete that had to be taken out because they were all honeycombed. Obviously the contractor just threw it in there and didn't vibrate it properly. There weren't enough inspectors around to check all of the concrete when the forms were removed to make sure that there were no imperfections. If that had been exclusively a Corps project, we would have asked for additional money and time and used the proper strength of concrete and made the corrections to the concrete facing problems along the way instead of leaving it. But we couldn't do that, because this wasn't our project. It's a shame because at the Corps, we knew that if it had been our job we would have gone in there and redesigned it and done it right the first time.

When the St. Lawrence Seaway system was opened for navigation in 1958, there was no suspicion at all about the integrity of the concrete. It took several years for the flaws to appear. In the early 1960s, the vertical lift gate seal on the Eisenhower had to be redone. Every winter after that, I went up there and made inspections. When I went up there with some of my colleagues in 1966, we finally decided that we had some big problems with the concrete on both locks. The biggest problem of all was the contractors did not have any reinforcing rods along the culverts. That was pretty much the way the Corps built things back then. When contractors poured massive amounts of concrete like they did on the Seaway dams and locks, no real analysis could be made that told them what the stresses were and where they needed to place steel. There was no water stopping that connected from top to bottom. Apparently equipment operators backfilled the walls with glacial till when they were still relatively new, so instead of the walls' withstanding the earth pressures we had designed for, they had to deal with three times that. So immediately upon backfilling, the upper landside corner of the culvert cracked right out to the back wall, which is a scary thing when you are in an earthquake zone. If an earthquake struck the area again like in the 1940s, I don't know what would have happened. I think people were doing a lot of praying for a few years. So many things went wrong even though the designers and most of the contractors did everything by the book. But the book and construction standards have changed since the 1950s.

In the winters of 1967–1968 and 1968–1969 the Corps made major repairs to both the Snell and Eisenhower Locks. Our workers covered and heated both locks and relined the culvert concrete and the laterals down below. Both locks need to be rebuilt with better concrete. Nobody really appreciated how bad the concrete was until the sixties when the scaling on the face of both locks started showing up. There are many reasons why the concrete failed, including the extreme weather and temperature variations; the height of the locks, which are over one hundred feet to obtain the forty-five-foot lift; and the one hundred feet of water pressure pushing on the concrete down at the bottom. It was really a combination of things that caused these imperfections and structural defects.

Somehow the House Civil Works Subcommittee got involved in the 1980s. The Corps sent me to Washington. The administrator of the St. Lawrence Seaway Development Corporation and his Canadian counterpart were in attendance. Both of them asked me a number of questions and I told them the truth. I didn't try to hide anything. I told them there were some major repairs that still needed to be done. The Corps construction section would have been glad to complete this work for them, if they gained the approval of the necessary funding. But everyone was fighting for money back in those days and the Corps budget couldn't fund a major rehabilitation project. Besides, the maintenance of the Seaway, locks, and dams was not our responsibility.

George Haineault

The power dam continued to be part of my life after the construction was over. In 1993 I purchased a motel and for three months every spring and fall when the energy consumption level was low representatives from Westinghouse, one of the companies that supplied the turbines for the Robert Saunders Dam, refurbished the generators. Hydro operating crews had to recircle them and, of course, at the same time rearmor them. That team would always take up six of my rooms and pay the nightly rate, which was a good deal for me. Then another guy came to stay with me for eighteen months to deal with the damage to the concrete caused by algae. Similar to a small clump of weeds that grows through a crack in a driveway, the algae grew through the cracks in the concrete on the Canadian side of the power dam and caused it to heave.

During the construction, the cement crews had left expansion joints around each cell to allow for contraction and expansion. But the algae grew beyond the limits of those expansion joints. So the contractors hired by Ontario Hydro took a cable with teeth in it and wrapped it around the dam. This cable rotated, cutting through the algae. A lot of these sections that were stripped had to be regrouted. The man who was staying at my motel found out that I had worked on the original construction of the power dam. He took me down into the lower observation deck because he wanted to know how different it looked from when it was first built. I spent a day and a half down there. It was interesting to see what it looked like under water and how it all functioned. When I was working on it, it was just a big wall of concrete. I think Hydro engineers deserve a lot of credit as they were able to solve the problem of the expansion algae. I have no doubt that the Saunders Dam will still be in operation in the next century.

6

Off the Project

THE 1950S HAD OFTEN BEEN PORTRAYED as the final stage of the development of the American consumer economy and the decade when people began to believe they could attain the good life. Members of all classes earned higher salaries, which equated to new purchasing power and a love of life. According to Douglas T. Miller and Marion Nowak, individuals spent one-sixth of their personal income on leisure activities.[1] The consumption of alcohol, illegal gambling, and partying skyrocketed among both the male and female members of the working, middle, and upper classes. Americans traveled to new places for vacations and to seek employment, purchased automobiles annually, and became first-time homeowners.

Federal government officials supported this new lifestyle in 1955 when they raised the minimum wage from $0.75 to $1.00. By the end of the decade, the average worker took home $5,000 a year, a 61 percent increase from 1950.[2] Citizens rushed to spend the money and catapulted the nation to a level of prosperity not seen since before the Great Depression. However, pockets of poverty still existed across the United States in rural areas and small towns where the lives and incomes of farmers and factory workers remained unchanged. During the St. Lawrence Seaway and Power Project in Massena and Cornwall, men and women who participated in this new culture of prosperity collided with residents of a town whose social and financial progress had stood still for decades.

Based on this insular attitude, many workers and their families described their tenure in Massena as trying to make the best out of a difficult situation. Harry H. McLean, a concrete engineer from Uhl, Hall, and Rich, discussed his and his co-workers' impressions of the area and their

25. Ex-president Harry Truman views the murals painted by his friend Thomas Hart Benton at the visitor's center at the main power dam. At the left is Robert Moses, chairman of the New York State Power Authority, and Charles Poletti, trustee of the New York State Power Authority and former governor of New York State. Courtesy of Alfred Mellett and PASNY.

interaction with local residents in a five-minute speech to the Massena Toastmasters Club in 1956. "During the time I have been in Massena, I have heard many newcomers talk of their impressions of the community. Nearly all who speak are critical; those who approve are silent. I believe in time to turn this about and give constructive criticism, to talk about how guests should treat their hosts. We came to Massena on September 15, 1954, and in a surprisingly short time we did know that Massena cared. Churches, clubs, organizations, but perhaps best of all our next-door neighbors showed an interest in us and helped us feel at home. I find it difficult to listen when some guests of Massena expound on the faults of Massena. He will never admit the advantages and benefits."[3] These words exposed the mixed experiences of workers and their families in Massena and Cornwall.

In 1956 Alan Emory, in an article for the *Watertown Daily Times,* praised investors, contractors, and lead agencies involved in the St. Lawrence Seaway and Power Project for garnering the area national exposure in a segment on *Wide Wide World.* He also highlighted how workers and tourists had infused money into area businesses. However, he bemoaned the increase in traffic and the effect workers and their families had on Massena's already dismal housing market, stating that trailer parks and run-down boarding houses and homes made the "shabby areas of town even shabbier." He also inferred that residents felt that their personal safety had been sacrificed along with many of their family homesteads and landmarks. Massenans locked their doors and no longer let their children out alone.[4] Welcoming editorial writers tried to overshadow the negative commentary. Many realized that the workers and their families were moving to the area regardless of whether the town wanted them to or not. In November 1955 Dick Peer, in his column "Peering at Massena," welcomed newcomers and said that the two best places in the world were Massena and their hometown. He added that he hoped that workers and their families would get to know local residents better by attending churches, lodges, and fraternal organization meetings, and by taking part in community affairs and helping the community become bigger and better.[5]

The initial contingent of Seaway workers and their families who arrived in Massena read the critical newspaper articles describing them as social deviants. Department of Justice officials warned the local and state police officers that, based on statistics from other major public works projects, the increasing population and their untamed lifestyles would cause an escalation in murders and other violent crimes as well as prostitution. Workers faced the difficult task of living in a community populated by fearful individuals historically known for their dislike of outsiders and for their love of small-town isolation. Many newcomers sought to make a positive impact on the community and prove their critics wrong.

One American worker, Joseph Marmo, created the youth hockey program in Massena and raised the funds for an indoor rink. Others established Little League baseball and a local chapter of the Toastmasters Club. In 1957 the President's Council on Youth Fitness 1957 caused the development of Little Leagues and organized sports for children across the country. A new generation of men also took up golfing as a hobby based on the game's newfound popularity, attributed to the arrival on the professional tour of players like Ben Hogan.[6] The workers who constructed the Seaway and Power Project also knew the value of a good education given that many had earned the first college degree in their family. They along with their wives were active members of the Parent-Teacher Association and forced local school board officials to erect new buildings to accommodate their children's needs in an updated environment. Many workers tried to put as much back into the community as they had taken out. They gave a culturally stagnant town a much-needed breath of fresh air. The Seaway workers in many ways left an unforgettable impression on the residents of Massena.

Joe Marmo

When I first arrived in Massena, I was surprised to read what have become known as the very notorious antiworker articles in the *Massena Observer*. The reporters wrote some nasty stories about us and I was furious about it. They didn't even know us and they were already telling people to expect the worst. The articles said that the riffraff was here and that residents were just going to have to live with them. They predicted that the number of violent crimes and rapes would escalate. I think the hardest thing was that these comments started us off on the wrong foot in terms of our relationship with local residents. Many people believe everything they read. Mothers and fathers in Massena thought they had to protect their daughters from these strangers. But I just ignored it because I knew I was not like that. When they found out that people like me started hockey programs and my co-workers established football and baseball leagues for the kids and went to church on Sunday, I think many of the local government officials and the newspaper editor felt a bit guilty.

When I arrived in Massena, I had recently graduated from college, where I had completed a successful hockey career. So when I started working on the Seaway and Power Project, I still had my hockey stick. Not long after I moved to the area, I jumped on a sheet of ice on an isolated part of the St. Lawrence River with my hockey stick and a puck. Some workers came out of a nearby shed that they were using as a temporary office and threw me off the ice. Being a cocky Italian, I said, "You have no right to do this. I will be back!" The next thing you know I started the youth ice hockey program in Massena, which was very successful. It really blossomed and I ended up raising the funds to build an indoor rink. The program gained such a great reputation that some of the players got accepted into college on hockey scholarships. I started my career working with kids up in Massena.

Jack Bryant

For centuries the residents of Massena lived in relative isolation. People thought differently, went to school differently, and worshipped differently than all of the construction people who started swarming in and they were unwilling to adapt to our needs or accept our different cultures. When we arrived in Massena in 1955 most of the local residents did not know what to make of us. I think most of the Uhl, Hall, and Rich employees and their families may have had an elitist attitude. Most of us were better educated than the construction stiffs. The shop owners traditionally had catered to the locals, who did not have very discerning taste. Most of the stores were open from nine to five Monday to Friday and a few had extended hours on Saturday until one. But on the Seaway workers labored on three shifts and when they got off they wanted to be able to buy things. They had money to spend. About halfway through the project a local developer built a little strip mall where one shopkeeper kept his establishment open twenty-four hours a day.

Many of the workers, including myself, had a hard time shopping for shoes and clothes. Based on the French Canadian origin of many of the people in Massena, most were generally shorter than your average man. It was also a one-industry town where workers made low wages, so the quality of the clothes and other merchandise that store owners stocked was poor. The largest suit size that the owner of the local men's store had on his rack was a thirty-eight medium, while the shoe store owner considered anything over a size nine and one-half to be a special order.

The muscular and hearty carpenters and concrete workers on the project needed forty-four-long jackets and size fourteen shoes. The tailors and the shoe salesmen couldn't believe huge men like us existed. We also had a higher standard in terms of alcohol. The owner of Romeo's Liquor Store usually sold patrons a bottle of cheap whiskey once a week. We wanted to buy cases of the good stuff. When we had a party, the host would provide the best liquor and wine available. The atmosphere was almost frontier-like, you might say.

The fact that Massena residents couldn't swim and distrusted outsiders amazed me. Men and women would stand along the banks of the Grasse and Raquette Rivers fishing and if they fell in they drowned. It was as simple as that. Therefore there was a lot of resistance when the Massena school board held a vote for a new high school which included a swimming pool. It was just a strange atmosphere. Most people were very nice; however, they were a little spooked by us. For example, if our kids came down with an illness, my wife and I could get the local doctor to make a house visit, but he always wanted to be paid in cash on the spot. We were strangers to him and he thought we might leave town without paying our bill. Generally I guess our people kept pretty much to themselves.

I enjoyed talking with the Indians and trying to socialize with them, but there was always a veil. I remember seeing a box lacrosse match at Rooseveltown between the Indian team and one from Niagara Falls. It was a revelation! The players beat each other with their sticks. It was worse than a hockey game. My friends and I also went across to Cornwall to a little old rink to watch hockey. The building was equipped with a heating pipe that ran underneath the bleachers. There was heat for the first period and than you were on your own. It was just a little place, but the players were vigorous and the fans were real hockey lovers.

Roy Simonds

When I got there, Massena was not a large town. It had a population of only about fifteen thousand, so the community found it difficult to absorb that very intense population increase, plus they thought their town was being invaded by the scum of the earth. Residents eventually found out that the people who came in for the project really dived in and did a lot for the town. When we first got there they would hardly wait on me in the stores. If the clerk didn't recognize me as being one of the natives, he took his time about waiting on me. Editorials came out in the paper that told residents to keep their children inside because the riffraff was coming to town. However, even with this negativity, it was like a vacation for many of the families, including my own.

Jim Cotter

I didn't have much contact with the local people off-site. Before our arrival, Massena residents and newspaper reporters referred to us as "interlopers." That was the term used to describe transient construction workers like me, and it was not meant to be a compliment. They argued that we were there to invade their town for a short period of time and then leave. In many respects that was an accurate description. But we did make a difference and helped pass a bond for the construction of new schools.

My main interaction with local businessmen involved my one-year term as secretary for the Toastmasters Club. Harrel Yaden, who was a media spokesperson for the Power Authority of the State of New York, started the local chapter. Its membership included local businessmen as well as employees of the Power Authority and Uhl, Hall, and Rich. We met once a week for our regular speaking efforts. Other members of the group included the local physician, Doctor Ingram; the owner of the sporting goods store; and the editor of the local paper, Leonard Prince. He had moved to Massena sometime in the late 1920s because he read that they were going to build the Seaway and he wanted to be there on the ground floor.

My family and I also enjoyed exploring the area and shopping in surrounding towns on both sides of the border. We went to tailor shops in Canada and got custom-made clothing because after World War II the Canadian government had encouraged skilled craftsmen from Europe to migrate to Canada. The stores were filled with high-quality items, including furniture at good prices, that weren't available on the American side. Massena had some great hardware stores where I found horseshoes and antiques that I had never seen west of the Mississippi. My wife and I also

26. Members of the Massena Toastmasters Club, including Jim Cotter; Leonard Prince, editor of the *Massena Observer*; and Harrel Yaden, media spokesman for PASNY are pictured. Courtesy of Jim Cotter.

were very interested in the historical side of things and it gave us time to visit historic landmarks that we could never do out West. I was always working on projects that were in the middle of nowhere and far from established towns and cities. So it was a highlight in my career and in my life in general.

Keith Henry

Contractors brought in almost all of their skilled workers, particularly at the Massena sites. The people in Massena thought they were going to get all sorts of jobs and when they didn't, resentment surfaced. Instead, nearly four thousand people moved into Massena. These people came in from jobs out West, including the Colorado River. They just followed the construction around and they came in and took all the jobs. I remember in the little town of Waddington, some of the local people bought restaurants and the bars and made a fortune. Other entrepreneurs came in from outside and purchased existing businesses because they were being sold at cheap prices. The workers made the area lively, all right. Coming from the West, I didn't think it was that wild. As most of the workers owned cars, they filled the streets with traffic and they and their wives packed the stores and taverns. There were an awful lot of people around. Some of the other industries that often turned up on large construction projects, like gambling and prostitution, also took up residence in the bars.

Massena merchants and single women were more welcoming of the new arrivals. Workers and engineers who earned wages above the national average bought cars, started savings accounts, and illustrated the consumerism of the 1950s. Massena storeowners, restaurateurs, and bar owners embraced the Seaway employees because these men and their families purchased goods and services which increased their profits. Local women were also very hospitable to Seaway workers. Several single Massena and Cornwall women married Seaway workers and embarked on a transient lifestyle. Many relished the opportunity to marry someone with an exciting career and promising future that could help them escape their hometowns and families. Workers and their families also enjoyed making trips to the nearby Adirondacks, attended church, and often stopped in a bar or restaurant to discuss the events of the day with their co-workers.

Thomas Rink

I had a great time when I was in Massena. I bowled in three different leagues with all the officers in the laborers' union like Sam Agati and his brother and Bob Ashley. I also played softball with Bob. The union leaders ran the whole show. I got into the union pretty deep through bowling. I was pretty happy back then. I was young. I used to stop in the bars and shoot the bull. That type of culture is gone. It was just like the American West. Everybody drank and drove back then. No animosity existed between the local workers and the transplants like me. There was plenty of work. Everybody was buying new cars. Some people overbought. Others saved their money. The cost of living was also a lot lower and a dollar was worth a dollar. I was only making $1.60 an hour back then; $53.32 was the amount of my first paycheck, but my money went further. I could buy ten gallons of gas with a dollar. When I got all done paying my bills, I still had some money left. I also could have bought a new car for $2,000 or $3,000. Instead, I purchased a used car that was very inexpensive and in good shape, but most men bought a new car every year. It was great times. I also played softball in the summertime and I went to the Adirondacks. I just loved Lake Placid and Cranberry Lake. I hunted for white rabbit and went up to Montreal and Ottawa a couple times and I did a lot of fishing with my brother on the St. Lawrence. I had an enjoyable three years up there.

The local people from Massena were very welcoming and very friendly. If I went into one of the restaurants, they talked to me. I just felt at home. When I was in the service at Fort Knox and I went into town they didn't want to have anything to do with me. They thought I was trouble. The people up there in the North were very nice to talk to. I am sure other people have given a different impression. When I got into bowling 90 percent of the people involved were locals. They were just fun people to play cards with or have a drink with or go to their house. They accepted me. Many owned their own businesses and restaurants and it just thrived up there.

Kenneth Hallock

The presence of all of these new residents in Massena increased business for shops and taverns and made interaction with the local residents necessary. This added a whole different dimension to the project. My interaction with the community was relatively limited, mostly with my neighbors and storekeepers. I felt welcomed all the time and I was very happy with the way people treated me and my co-workers.

Ray Singleton

I was always the kind of person who got along with everyone every place I went. I met my wife during an operating engineers' strike. I had two weeks off and I saw her every night. She was the sister of one of my co-workers, Willie Rutley. That is when I went to visit all of her people. I was not the only worker to marry a local girl from Cornwall or Massena. A few other men met girls in bars or through co-workers. I wasn't even thinking about getting married. I was twenty-seven years old and was enjoying the single life until I started dating Melba. She was a year older than me and I think she took advantage of my youth. She was one of those northern, sophisticated, professional women who took advantage of a dumb old southerner.

A lot of outsiders who came to Massena did not like the way that the northerners acted or treated them, but I never found it that way. I didn't have any problems dealing with the locals in restaurants or in stores because I had traveled extensively for my work and adapted well. I know a lot of the natives in Massena didn't like us operating engineers because they thought if they could drive a tractor, they could handle a crane or bulldozer. The contractors wanted skilled workers to operate the large machinery, so a lot of the operators were transferred in from Syracuse. To learn how to run that stuff was not a cut-and-dry situation.

Business owners' sales rose because there was an influx of people with money who wanted to spend it. Everybody was doing really well. The biggest problem associated with the large number of workers and their families temporarily moving to Massena was that it caused the mayor and the school board members to raise school and property taxes to fund the extra services like new schools and roads. Many retired people had to sell their homes because they couldn't afford the upkeep and the maintenance costs

27. Melba and Ray Singleton on their wedding day, November 10, 1956. Courtesy of Melba and Ray Singleton.

along with the new taxes. Some made extra money by renting out rooms to workers or creating trailer parks on their vacant land.

Roderick Nicklaw

I met a lot of wives whose husbands worked on the project and they weren't exactly faithful or spending their nights at home alone. When their husbands worked the midnight shift, they weren't at home knitting. Some of their husbands knew and didn't mind that someone was keeping them company. Others were more discreet and had a boyfriend on the side who worked the opposite shift of their spouse on either the project or at Alcoa. Sometimes the husbands found out and the affair came to an end very quickly. This happens on every construction project and in many areas with large manufacturing plants. These cases may have been few and far between, but I knew a lot of women that ran around.

Barbara Brennan Taylor

I never heard anything about prostitution. Among the local women, there was a lot of reluctance about starting a relationship with one of the workers, myself included. Most of these men went from one construction site to the next, and if you didn't embrace that kind of lifestyle, you didn't really want to get involved. In those days everything was not as free and open as it is today. Pregnancy was a disgrace if you were not married. So I don't think the lifestyle was as wild as some people have tried to portray it. My female friends and I had a wonderful time, but there was still a bit of caution. It was just different. I lived at home with my mother and she often wondered where I was. She probably had one eye and her ears open when I was out at night, but she was very tolerant.

Cornwall residents were no more welcoming of Seaway workers than Massenans. Historically the town had witnessed a diversification of its population, initially with the recruitment of foreign workers to complete the Cornwall Canal in 1843 and French-Canadian workers to serve as mill workers at the first textile plants in 1867. Instead of making the existing population more accepting of outsiders, the influxes increased their resistance and jaded their opinion of new arrivals. In both cases, these individuals drank heavily, committed crimes, and seemed to present a moral threat to the town's social fabric. In 1954, the distrust of new town residents remained and Cornwall residents' greatest fears became reality. Unlike workers on the American side, concrete workers, equipment operators, and carpenters did not move on to the next job site after the project's completion. Instead they liked the area so much that they remained in Cornwall to raise their families. Some established businesses, while others found work at the newly created headquarters of the Seaway Authority.

Les Cruickshank

I lived and owned a business in Cornwall for a long time before I was accepted. Even after the Seaway, I had a competitor in town and he used to say, "The Cruickshanks are just camping out." It was not a warm and welcoming atmosphere when the workers arrived. I wouldn't say they were trying to run you out of town or anything like that. It was overwhelming for the people who lived here. There was a pile of transients that moved in and they were strange to them. All these new people arrived almost overnight with their families and different cultures. Cornwall residents were happy to have all of these new jobs, but with all these new people coming in they no longer knew their neighbors. The small-town atmosphere and security temporarily vanished. There was some justification for their suspicious attitude. Most of them won't admit that they felt that way about the short-term changes that took place during the project.

Arthur Murphy

The city of Cornwall was different during the construction. Normally eight cars in Cornwall at a traffic light was a traffic jam. The increase in traffic caused many funny things to happen. As men tried to get to work, their wives drove to the stores, and the school bus drivers took kids to school, traffic on the main roads barely moved. The public works department put up lights at some of the intersections and policemen at others. The town engineers had to redesign the infrastructure of the community to accommodate all the traffic. The board of education also raised taxes to build a new high school. There was a fair amount of improvements that probably would not have happened without the Seaway and Power Project. It certainly meant a lot of business for the store and bar owners.

Economically it just boosted the area because a lot of construction workers lived here and spent their hard-earned dollars in town. Businesses and the economy flourished for about five years. But of course after the flooding and the power house was built, the construction workers moved on to other hydraulic and nuclear sites in the province. These workers were all transient types and took their families with them from one project to another. In my case, I married the only daughter of one of Hydro's top administrators. She was a cheerleader for our high school. A lot of workers on the Canadian side, especially the engineers and the supervisors, were older, so I went to school with many of their children.

Arnold Shane

If you were working on the dam and Seaway, you were making good money. There weren't many places to spend your money. About five thousand people moved in to Cornwall and really made an economic impact for a few years at a time when many locals were not employed. I was just out of school at the time and started out at the lowest level of pay, and when I left I was concrete inspector and was making better pay. Most workers went bar-hopping every night. I didn't. I played on Hydro's hockey and basketball teams. But there was a team for almost every sport. It was widely known that Hydro actively recruited workers for their large projects who were not only skilled and experienced tradesmen, but avid athletes. It created a kind of camaraderie outside the workplace. They also had a well-organized dance at Christmas time. It made me feel loyal. Bill Goodrich was in personnel and he liked his sports too. If one of our teams was not doing well, he would search through the Ontario Hydro payroll rosters to see if there were any former athletes employed at other sites. Then he called the personnel director at that location and had that individual transferred to Cornwall. Many times the worker would be unaware of the reason for his transfer. Even if he was not a skilled craftsman or there was not a shortage in his particular specialty, the foreman found work for him.

Bill Goodrich

In terms of interaction with the Americans as well as among workers on our side of the border, I coordinated numerous sports activities. The congeniality among workers and administrators in the United States and Canada is something that is hard to explain, but was key to the project's successful and timely completion. I organized a nine-hole golf tournament twice a month on Friday afternoon that pitted the Canadians against the Americans. It became so popular that we virtually had to shut down the whole project on those afternoons because all of the engineers and supervisors were on the golf course. Even if someone didn't play golf, they sat at the bar.

Communication and getting to know each other was very important. We all were working toward the same goal and needed to understand each others' problems and find solutions. One area where we interacted with the local residents, even after some resistance, was in the sports arena. At the beginning of the project, one of my colleagues and I wanted to play for the north end fastball team in Cornwall, but we were turned down. So I decided, along with my buddy who looked like a gorilla and was my secretary's husband, to go down to city hall and convince the town leaders to let us organize our own team of players from Hydro and participate in the regional league. The following night we went to Montreal and hired the two best fastball players in Canada. I put them on a survey crew. I also organized teams for other sports, including hockey. If a worker was interested, I started a team. Finally, I actively recruited workers from other Hydro sites to bolster the talent on our teams and to ensure a winning season. It was enjoyable, that's all I can tell you. The people I worked with were fun-loving.

According to David Halberstram, author of The Fifties, *at the core of the American vision of a better life was home ownership. It embodied the new American dream and strengthened the family structure.[7] Many couples who had rented apartments in the past purchased homes or trailers in the 1950s. Most paid $5,000 for their first home, which was equal to two years of the average family income. The mass construction techniques developed by Bill Levitt made these houses more affordable. These units designed for young families contained four and one-half rooms. Workers assembled many of the sections of the house at a factory and then truckers transported them to lots in new developments. This method required fewer on-site skilled workers, cutting contractors' costs and resulting in lower asking prices. Many members of the middle class reacted with pride and excitement over their ability to become the first homeowners in their families.[8] In Massena, engineers employed by*

Uhl, Hall, and Rich and their families lived in Levitt-inspired homes in the Buckeye Development. Factories in the Buckeye State manufactured sections of each of these homes and shipped them in pieces to be assembled on the newly created cul-de-sac. For single workers and the remaining families, trailers, renting rooms in a motel or existing residence, or leasing an entire home seemed to be the best options.

Robert Hampton

Uhl, Hall, and Rich had built subsidized homes for their engineers and their families. Most of us brought our families and belonged to the PTA. Our kids joined the scout troops and there was a very friendly atmosphere. Many people talked about the rough towns where they had lived before. Massena was a much different atmosphere. Most of the workers and their families went to church and everybody got integrated. Our main interaction and support system came from our immediate neighbors. In fact most of the people in our community had kids the same age. They would get outside and play and it was understood that if the kids went to a different house, the mother from that house would make sure that they were okay and didn't get snow in their boots and freeze. We also all had dogs. All I had to do if I wanted to go find my kids was to look at which house my dog was sitting in front of. I would say that we were kind of part of a community. In some of the projects out West where they built a village they would have automatically had clubs and social events and it was the same in Massena. It was quite an experience. My wife was always going off to do arts and crafts and or to have coffee and playing cards. My wife and I always knew that we could count on our neighbors. These people were all young married couples like ourselves, so we had a lot in common.

Jim Cotter

In June 1955 workers and their families descended on the Massena area by the thousands and I was lucky to find a one-room shelter over a garage. Fortunately it did have running water, a bathroom, stove, and beds. Since my daughter was only five weeks old, the washing of diapers and making formula were high on the list of "musts" each day. Twine strung between the rafters became the clothesline for drying diapers. That added to the humidity and discomfort of the ninety-degree daily high in the unvented attic. But such is what makes fond memories. My wife and I patiently awaited the completion of our assigned house in the Buckeye Village. Our house was on Middlebury next to the back door of the creamery. How handy that was with small milk-drinking children! Soon we were surrounded by neighbors and immersed in a community spirit. I was from a smaller town than Massena, so it wasn't a shock for me.

When I worked in Massena, it was my first occasion to live in a development with my co-workers. At the Hungry Horse project, the only housing that was available was a dormitory constructed by the Bureau of Reclamation. There was no housing provided for families, but on the Seaway project because my family and I were living in a development surrounded by an unaccepting native population, our social life, by and large, involved our neighbors. It isn't that we all got along like lovey-dovey, but I still made some very close friends. On other jobs I worked on there was little camaraderie among workers. So Massena was the first place I lived in a development where all the people were construction people and had something in common.

The homes were prefabricated and made in Ohio, the Buckeye State. Uhl, Hall, and Rich purchased and erected these homes for their engineers. The Seaway and Power Authority had not constructed any worker's housing and my employer feared that many of us would not relocate without adequate accommodations. It was an enticement because the company paid 50 percent of our rent. We were all brought together simultaneously in July 1955. We were thrown together under the same set of circumstances for a long period of time with the same work objective, goals, and family responsibilities. We were all far away from our family and friends and living in a new region of the country. Many of us were in the same stages of our lives, raising young children, and found that we had a lot to talk about. I think it all helped us survive the rigors of the project and encouraged all of us to remain in the area until the project was completed.

Joe Marmo

For eight years I lived at 46 Woodlawn, which was a private home right across from the Woodlawn Hotel and Bar. The reason my wife and I lived in an existing neighborhood was that we arrived before Uhl, Hall, and

Rich had completed the Buckeye Development. I suppose I could have moved in the latter years, but I liked living separately from my co-workers. I did get involved with the native Massenans through golf. I became good friends with the owner of Charlie's Tavern, who, like myself, had been a tremendous high school athlete. We took up golf at about the same time and we spent a lot of time at the golf course. I don't think the adults made a lot of friends with members of the existing community, but our kids made friends with locals their age.

Kenneth Hallock

Literally thousands of workers and their families converged on Massena, which had a population of under fifteen thousand. It is not surprising that housing was very difficult to find. I had a problem locating a place to live, but fortunately my boss, Tom Airis, was buying a house when I arrived in Massena. He had been renting a house for a few months from a lady that lived in Washington, D.C. So he made arrangements for me to take over his lease. This project also stressed the infrastructure, businesses, and schools in the area more than others I had worked on because the sites were right in the middle of a town. Most of the projects I had worked on out West were far away from civilization, so the contractors and lead agencies just built workers' camps with dormitories and cafeterias. Most of the men I knew brought their families with them for the first time.

Bill Spriggs

When I first arrived in Massena, I rented a room in the middle of town. But then I got a unique opportunity to buy a small house. Being a veteran qualified me to purchase a surplus home that was built by the federal government in the area during World War II. The government had constructed and operated an aluminum processing plant next to the existing Alcoa facility in the late 1940s. When the war was over, the government put the houses up for bid to veterans only. So I bought practically a new two-story house at a very low price. I felt sorry for the other workers because there wasn't much decent housing available, and PASNY and the Seaway Authority did not build any new apartments or homes. Many families had no other option but to live in one of the many trailer parks.

Roderick Nicklaw

Most of the local workers who lived in nearby towns like Waddington, Potsdam, or Norwood drove to work every day. They owned houses in the area and it was cheaper for them to pay the cost of gas than to move to Massena. The workers who moved to the area temporarily had a harder time finding places to stay. Some of them rented rooms in private homes or in one of the local motels. As a way to cut costs, four or five guys would get together and rent a whole house and divide up the expenses. A lot of them were married and they were trying to conserve a good portion of their wages to send home to their wife and kids. Others who came with their families lived in trailers and mobile homes. This seemed to be the easiest way for the professional transient workers to save housing costs from job to job. I don't know if any of them owned a stationary home.

James Romano

When I first arrived in Massena, I stayed in a house in Norwood because that was where the gravel pits were located. The lady who owned the house gave me and one of my drivers the room for ten bucks apiece. The first morning I was there I went into the bathroom and there were three faucets. I knew one was hot, one was cold, but I didn't know what the third one was for. I found out later that it was for the rainwater that they had collected from the roof. So I was brushing my teeth with the septic tank water. I thought that the water was stinky. I just thought that it was like the sulfur water that people drank up there to stay healthy that was green and smelled like rotten eggs. Later on in the project, I stayed in the Excelsior Motel in Massena. I shared a room with another guy and we only got one towel a week.

Alfred Mellett

As a bachelor, I got to pick where I wanted to live in Massena. I liked to be somewhat removed from the other workers and my work schedule was unpredictable. I often photographed the various construction sites at all hours of the day and night, which made renting a room in a private home difficult. I also could not share a room with another worker because I never knew when I would be working and when I would be sleeping. So I rented a well-outfitted trailer down on the shore of the St. Lawrence River. The biggest problem was that it got terribly cold in there in the winter and extremely hot in the summer. No matter how much bedding and blankets I had, I still couldn't keep warm because my body was bringing in the cold air through the mattress. So I took a sheet of plastic and laid it underneath the mattress. I kept nice and warm from then on. The view of the river was spectacular, though, and I was often invited to dinner or to parties at the engineers' homes. Their wives often had me over for coffee and loved to hear my stories about the many dignitaries I had photographed who had visited the area. Being single definitely had its advantages.

Ambrose Andre

The typical room workers could rent from an existing homeowner was $10 a week and they were sort of like dormitories. You had a bed and a dresser and you shared a bathroom. With shift work, people were coming and going. It was not a place where you would sit down and read a book. Television was pretty new at that time. You worked or went to a bar. One of the rooms that I had was with a Canadian couple and I am quite sure that they had just walked over the bridge and never went home. They were friendly, but they didn't want me to get to know them because they didn't know if they could trust me. I am quite certain that they were illegal. They were the only ones that I wouldn't call friendly.

Floyd Grant

In and around Massena, farmers and other landowners created trailer parks. People like my sister, my parents, and many of their neighbors rented out their land to workers to put their trailers on. Prior to renting out these plots, the town board required owners to put in septic systems as well as big wells. They all liked the families they rented to and got along well with them. It was also a very welcome addition to their income. A lot of men also brought their families, so local schools were overloaded at first. A few years after the project started, they built a big school in Louisville. Before, the residents of outlying areas had their kids bussed into Madrid or Massena. In terms of crime, I don't think there was a murder in the area between 1954 and 1959. As more workers came and were hanging out at the bars, there were a few more fights, but nothing serious. I used to go out dancing on Saturday night and there was never any real trouble.

Living in Cornwall offered a unique experience for many of the workers and their families. In the past, Ontario Hydro projects had been located on waterways far from established towns and cities. The overseeing agencies had constructed dormitories and cafeterias, making it impossible for workers to bring their families. As the St. Lawrence Seaway and Power Project was based in the heart of Cornwall, schooling and a variety of housing options were available. However, because Cornwall had never before experienced such an increase in population, it was not surprising that housing was scarce for the workers. Ontario Hydro built houses in existing neighborhoods for some of their top officials, while the remaining carpenters, laborers, and electricians lived in crude huts or trailers, boardinghouses, or apartments. The local workers already owned homes either in town or in the surrounding area.

Garry Moore

Beginning in 1954 workers and their families invaded Cornwall. Almost every day was a shock because more and more people and machines came in. Businesses started to make more money, and property owners rented their homes and apartments. It sure filled up the town. Residents with large houses and extra rooms rented them to workers. Everybody's attic had a body or two sleeping in it. There were five thousand people working on the hydro project and the Seaway and they had no place to stay. Several workers shared the rent on one room and slept on a rotating basis. Of course they made sure they worked on different shifts.

Arnold Shane

The biggest problem was getting all the workers housed. Six months into the project, there were no vacant homes or apartments available in Cornwall. Everything was owned or rented. Hydro built a lot of houses in the Riverdale area for their executives. There was a small camp erected near the site to feed and house workers who were new to the area or looking for work elsewhere. This was meant to be temporary lodging until a worker could find more permanent housing. It certainly wasn't large enough to accommodate five thousand people. If I am not mistaken, they were given a hut until they could find accommodations of their own. A lot of them initially came alone and then sent for their families. On other projects, Ontario Hydro officials set up camps on-site to feed and house workers for the duration of the project.

George Haineault

When the St. Lawrence Power Dam construction started in 1954, the housing market was already tight in Cornwall. Most of the permanent residents either owned their homes or had long-term leases on their apartments. So temporary accommodation in and around Cornwall, barring a few motels and boardinghouses, was almost nonexistent. Ontario Hydro officials only ordered their construction division to build a few homes for their top engineers and project managers. The rest of the workforce was left on their own. Some local residents who owned one of the big older homes made a fortune renting out their extra rooms, sometimes to more than one worker. Many charged higher rent than usual because they knew workers needed somewhere to stay, and many of them were making a lot of money. Many of these places were one hundred years old and barely standing up, but men still paid a premium price to live there.

John Moss

In terms of accommodations, I was one of the lucky ones. I already owned a home, but many people rented homes and apartments. It seemed that all of a sudden there was an influx of people with nowhere to live. There were boardinghouses for many of the single workers. Many workers shared a room with someone on another shift. Local residents with large homes or extra bedrooms also took in boarders. Hydro also built some houses for their administrators. Everyone fit in somehow.

Les Cruickshank

Ontario Hydro employed thousands of men on the project who were all making lots of money. The stores were booming. Initially I couldn't find a place to stay. When I decided to get married, I looked around for a decent apartment or house to rent and I couldn't find anything suitable. I was also competing for housing with residents whose homes were in the process of being moved or had been destroyed to make room for the flooding. Some of my co-workers rented rooms in some of the old hotels. Many of them weren't updated and they were pretty rough places. So I bought a mobile home and put it in a motor home park. I lived there with my wife for four years. Farmers and private landowners set up three trailer parks in Cornwall and one in Morrisburg, Ontario. Ontario Hydro had dormitories and trailer parks set up in Niagara Falls, but they didn't construct any when they came to Cornwall.

Bart Whitten

When I was a concrete inspector on the power dam, I lived at two boardinghouses because I was single. The owners were friendly towards boarders and I kept in contact with them after the job was done. Unlike on previous projects, Ontario Hydro only built one camp for workers where food was provided because there was no demand for it. Local builders had constructed apartment buildings in anticipation of the projects and there were a few trailer parks scattered around the three-county area.

In terms of the community, Cornwall was different than Niagara Falls. The people were friendlier. One hotel owner ran a booming business with prostitutes. People were always in the bars. There was little trouble because workers were afraid of losing their job. You could get fired if you got into a scrap. If one of the workers got in trouble off-site, Hydro officials were always interested in whether it had something to do with the job or not.

Bill Goodrich

Workers lived all over the place, in homes, apartments, and trailers stretching from Prescott to Cardinal, Ontario. Accommodations were virtually nonexistent when I arrived to take my position as personnel director in 1954. My boss, who was the director of employee relations and eventually became the director of industrial relations for all of Ontario Hydro, first lived in a motel and then in a small apartment. He was a married man with two children and weighed almost 280 pounds. He wasn't very happy. The Ontario Hydro construction division began building houses for the administrators soon after that. I was lucky enough to get one of those homes. When new workers came in, it was part of my job to help them find housing, but it wasn't easy. Many residents of Cornwall were very distrustful of outsiders and didn't want them taking over their neighborhoods. They also weren't happy that the housing scarcity caused an escalation in the rent and made it difficult for permanent residents to find places to live. Most men and their families found somewhere to live and, once the project was over with, everything returned to normal.

Donald Rankin

Before I moved my family to Cornwall, I came down from Niagara Falls on an exploratory basis to look for a house. A number of Hydro people, similar to me, lived in houses fairly close to the project in the Riverdale area, which had been constructed by Ontario Hydro. Eight families resided in that area, including the project engineer in the very early stages, Bill Hull. The eventual project manager, Gordon Mitchell, built a second house in the area, but the rest of us just rented the houses from Ontario Hydro. I suppose it was hard for common workers to find houses. I had come off the High Portage project, which was a remote project where we all lived in dormitories. On the Niagara Falls project, we were also provided with accommodations. But the eight thousand workers involved on the Canadian side of the St. Lawrence Seaway and Power Project were left on their own to find room and board. It was easier for them in Cornwall because we were more accepted by members of the community than on the Niagara project. I always felt more comfortable in Cornwall. Niagara Falls is accustomed to a lot of tourist people and they perhaps were not as welcoming to strangers as Cornwall residents. The locals here welcomed many of the workers into their homes as boarders and I don't think that would have been possible or feasible on the other projects I was involved with during my career.

Keith Henry

I lived in Cornwall in two different Ontario Hydro houses. Ontario Hydro built ten very nice small, comfortable, one-story brick houses. I got one of them because I was coming in from outside and had nowhere to go. I paid $90 a month for rent. Marg and I, with our daughters, Jeanne and Kate, moved to Cornwall in February 1956. We had sold our house in Toronto and shipped our furniture to be unloaded in the house at 103 Robertson Avenue in the west end of Cornwall. I made sure the movers put things in the right rooms and got the beds set up so we'd have a place to sleep. We loaded our two-year-old and six-year-old into the car with a lot of luggage and our budgie, carefully covered to avoid pneumonia. Marg was about three months pregnant but in great shape, and we were looking forward to a new world. We made the drive in six hours and got to Cornwall to find the house nice and warm and everything just as I had left it—a mess, but ready for unpacking and looking homey already. We bought a brand new spinet piano and it was delivered and put into its place of honor at the same time as everything else. With a brand new home and a great family and a very challenging job that I was already fully familiar with, I figured the present was great and the future looked pretty damned rosy too.

Drinking was part of the American and Canadian social fabric in the 1950s. Men and women on both sides of the border ushered in a new level of consumption and the acceptance of alcohol. Many married and single men and women enjoyed parties, dances, and any events that involved drinking. During this time hard liquor also gained wide acceptance and was legalized to be served in bars. In the United States Americans drank 235 million gallons of liquor in 1960, versus 190 million gallons in 1950. Beer consumption also rose from twelve gallons to seventeen gallons per capita. Cocktail parties became popular among the middle class. Restaurants and lounges began to offer martinis and manhattans. Savoring a scotch, whiskey, or mixed drink when they arrived home became a traditional way for men to unwind. Many executives also conducted business meetings over drinks. Young men considered their first visit to a barroom as a key step toward becoming a man. As Robert H. Young and Nancy K. Young stated, "In The Catcher in the Rye *even the youthful Holden Caulfield visits a cocktail lounge because he knows drinking signifies an important rite of passage in America."[9]*

During the St. Lawrence Seaway and Power Project alcohol played a role in the leisure activities of all classes of workers from managers to laborers. A variety of options were available on both sides of the border. The upscale Cornwallis Hotel in Cornwall and the Village Inn in Massena sold mixed drinks and three-course meals to engineers and visiting dignitaries. For the engineers and their wives, cocktail parties filled their weekend social schedules. Most of the single workers

enjoyed the bar scene in Massena. The beer joints were full every night and every
weekend. Some men worked a ten-hour shift, went out and partied for ten hours,
and then went back to the project site and worked another full day. Having a beer
with one's concrete crew was part of the after-hours protocol even for those who
were underage. Workers often stopped at the local bar to rehash the events of the
day before going home. Others saw it as a way to try and meet a future wife.

Garry Moore

Cornwall was like a Wild West boom town during the gold rush. The
men on the project worked hard and played harder. They loved to drink,
gamble, and meet the local women. In the 1950s, Canada had stringent
alcohol consumption laws. Workers had to purchase hard liquor from
other sources. You couldn't buy hard liquor at a bar, only beer. Drinking
establishments also had separate rooms for men and ladies with escorts.
There was also another room to drink beer in. Another law that was carried over from World War II was that a patron could only have one pint
of beer or two glasses of draft on the table at any one time. Tavern owners also had to close down the bar for an hour from either six to seven or
seven to eight. Everybody was supposed to go home. I just went outside
and smoked or got a hamburger and went back in when the hour was
over. The legal drinking age was twenty-one in 1954, but I went into my
first bar when I was seventeen.

George Haineault

Sometimes when I worked the night shift, I wasn't always tired in the
morning so my friend, who weighed 225 pounds and was over six feet
tall, and I would go over to Massena, had a few beers, and got back home
around eleven o'clock. I tried to get some sleep before the next night shift,
but sometimes we stayed a little too long. One day we visited our American friends at a bar in Massena. They told us we weren't welcome and told
us where to find the border. We of course didn't listen and came back with
a few shiners. I think this happened a lot around the taverns when people
drank too much, but never on the job site. Most of these altercations didn't
result in any major injuries, just bruised egos. That is why my friend was
more suited for the transient construction life and stayed with Hydro after

the project was over. I don't think I ever won a fight with anyone. I always ended up on the wrong end of it. I think the guys on the other side were just as stressed as we were. If we took a punishing, it was not severe.

It was pretty typical to go and have a few quarts of beer after work. It was fashionable for some people to have their liquor with them at all times. I would go to one of the local bars with my friends and we would rebuild the events of the day while we shared a beer. It always helped that I could talk and sometimes laugh with the guys on my crew about general work problems and sometimes accidents that really were no laughing matter. But if I had not had that outlook, I would never have survived.

Arthur Murphy

When I was a young lad, my friends and I used to go party over in Massena with our wives and girlfriends. It was a very popular spot. There was a lot going on and the bars were always full. Massena residents opened more bars and places to go dancing than in Cornwall. Almost every night of the week you could watch a movie, find a dance band playing, or sit and have a quiet drink. In Cornwall our only choices were to go to the Cornwallis, which was very expensive, or one of the seedy neighborhood bars, neither of which was affordable or acceptable to bring a lady to if you wanted to impress her. Also, married workers didn't want to expose their wives to their drunk and loud co-workers. It allowed us to get away from many of our neighbors and meet new people. I knew a lot of residents and business people in Massena and I spent a lot of time there because I played semipro football for the Massena team.

Barbara Brennan Taylor

I basically remember the good times. As far as the women went, it was party time. Ontario Hydro organized a ball team and I remember the project engineer's secretary broke her arm because none of us were in shape to play softball the way we played it. So that caused a bit of a ruckus because she could no longer work because she was in a cast for six weeks and there was no insurance for us in the office.

I spent my time organizing parties, writing a column for the paper, running the bowling league, basically just having a great time. We used to have dances at the Cornwallis Hotel, really lovely dos. Of course everything was extremely well attended. The office staff and I pretty much financed the parties by charging admission. We didn't get any financial backing from any of the companies. They might have given us $100 to buy prizes, but all the attendees paid their own way. Many of the parties and outside activities were my ideas. Fortunately my boss was the office manager, but he was hardly ever there. I don't know where he was, in retrospect. Everybody was always so enthusiastic about any upcoming party. It seemed that we were always planning something from bowling banquets to dinners at the Village Inn to dances. Everyone seemed to be in good spirits.

Hubert Miron

Ontario Hydro had some great formal parties at the Cornwallis a few times a year. During the other months, my co-workers and I had to find other ways to spend our time off. All of us made good money. I took home $80 a week and some of my friends earned $250 a week. The electricians and the bulldozer operators used to spend all of their paychecks over the weekend and ask me to borrow some money on Monday. They were making a couple of hundred dollars and lost most of it gambling. Besides playing cards, they used to place wagers on their checks. The checks were numbered and they used to bet on who could make the best poker hand out of those numbers when they got their checks on Friday. It was winner takes all. It was fast. They lived from day to day. On big construction jobs there is always a lot of drinking and a lot of parties.

John Moss

You couldn't buy hard liquor at bars when the project started, but it became available about half way through. You could go to certain places that had licenses like the Cornwallis and buy mixed drinks, but most workers drank beer. The pubs also became more crowded. If you were used to going to a certain pub after work for a beer like most people did, it became almost impossible because all of a sudden it was full of people at all hours of the day. Because there were several shifts on the dam, there was always somebody there drinking. The pub owners did a fantastic business while that project was going on.

Keith Henry

I made many trips from Toronto to Cornwall by train. It left at 5:00 P.M. from Union Station and got into Cornwall about ten at night. We traveled first class, which meant that we had reserved seats in the parlor car so we could get together for a drink before dinner. The usual gang was many confreres from the Hydraulic Section and the Generation Department, who had to go down fairly often, but there were many others involved too, such as property and rehabilitation people, so I virtually never got on the train without finding at least half a dozen people I knew well enough to enjoy a few drinks with. The parlor cars were fitted out with single rows of big comfortable swivel chairs on each side of the aisle. This allowed us to sit opposite one another and face the center of the car. I could also turn around and face the big picture windows and watch the world go by. Those were the days when no liquor was served on trains, and that was just great because it meant I could bring along my own booze much more cheaply than paying bar prices for it. All I had to pay for was the mixers and ice. Our general practice was to bring along a twenty-six-ounce bottle of rum or rye and split it among four of us. We usually had three drinks apiece before dinner and were in fine fettle when it was time to eat.

I remember that every meal started off with a complimentary dish of big queen olives and celery in a bed of rice. Our bunch was very watchful on this point and woe betide the man who tried to get more than his share of either item—the approved fate was to put the remaining ice water over his lap, accompanied by effusive apologies. We always chose the second sitting of dinner, which was called at seven o'clock. This gave us time to finish our predinner drinks in comfort. While we normally finished by nine, the staff allowed us to linger over an extra pot of the best coffee I have ever known, till we arrived at the Cornwall station.

David Flewelling

When I started working on the project, I was nineteen going on twenty. At the end of my shift, I was pretty much on my own and didn't have a vehicle. I often walked downtown to one of the restaurants for my meals on the weekend. There was also a bar along the river that I went to quite often. But for the most part, I was working twelve hours a day, so I went back to the house where I was renting a room. The lady of the house always had supper ready. I sometimes would have a beer or two and then go back to my room and sack out. I always knew I had to get up and go to work the next day and I was trying to save money so I could go back to college, so I might not have been as wild as some of the other construction workers who had traveled to different projects their whole adult lives. Drinking was part of their normal routine and very much a part of the construction world. Most of them had cars and were out carousing most of the night and then going to work the next morning without much sleep. I think each morning some of them would have failed a breathalyzer test when they reported for work at the project sites.

Bill Spriggs

Massena was a very lively place during the project! I was so bogged down with work that I couldn't look around me to see what other people were doing during their free time. However, I saw what kind of shape they were in when they came to work the next morning. I am sure the bar and liquor store owners did a good business. While some of them may have drunk too much alcohol, many of them were very religious. All kinds of new people came to my church and many parishes built new buildings and new faiths organized congregations.

Ron Barkley

On any construction project, drinking goes without saying. It is just part of your life. During the Seaway project, life in the area was free and easygoing. Everybody was partying because we had lots of money. I think bar owners ended up making all the money. The watering holes were always very, very busy. All of the heavy construction workers, painters, ironworkers, and carpenters were heavy drinkers. There was nothing else to do. It was a way of relieving the pressures. In those days television was brand new and there was nothing else for people to do but socialize at the bar. It is not like today; people don't leave their houses today. We didn't have television or computers. We had a movie once a week and that was it. The town where I grew up and lived during the Seaway project, Waddington, New York, is so small that three hundred of the five hundred residents worked on the Seaway. So of course the local residents were friendly with the workers and their families. Basically, the whole town worked for the Seaway and the business owners happily welcomed new customers.

Most of the workers brought their families. However, the single guys and the men who left their families behind drank the most and were the wildest. A lot of the workers would go to outlying areas away from the river, like Norfolk, where the bars weren't so crowded. Massena and Potsdam also had good drinking places. The Canadians came over to our side to drink because the drinking age was eighteen at the time in New York State. Also, in Canada a single man could not walk into a bar with a lady. It was not mixed couples. It was segregated. One side was for women and one for men.

Thomas Sherry

It was just like living in a gold rush town. In Massena, everyone rented their rooms and apartments out to workers and the barrooms were just overflowing like the Old West. It was wild. A lot of fights took place among men who come in from out West and had worked on that type of construction before. I didn't hang out in Massena too much. I would go back and forth. I just heard stories and read about it in the newspaper. Shop, bar, and restaurant owners were making money hand over fist. A lot of the farmers rented land to trailer owners and sold their produce and meat and poultry to local businesses. It was like one big party. There was a lot of drinking and having a good time. The barrooms and restaurants were always crowded.

Roderick Nicklaw

I drank a few beers with other workers who weren't on my crew. I was sixteen, but when I walked into a bar with my Seaway clothes, the bartender always served me a beer. He figured I was eighteen because I was supposed to be that old to work on the project. I worked hard and I played hard. There were a lot of things going on. Hundreds of bars lined the streets in every town area because that is where most of the workers spent their money. Malone was the same way. There was a bar on every corner.

In Malone four or five places had bands every night, so I used to dance, party, raise hell with my friends, and have a good time. We didn't use drugs, but we drank beer and sometimes there were a few barroom fights. Some of the southerners were still fighting the Civil War. I think most everybody got along. You are always going to get people that you don't like. But people made a lot of money in Massena and the surrounding towns. The local residents and business owners had never seen money like that before. People bought new cars, new clothes, and had a good time, and I was a part of that. I was taking home one hundred dollars a week. I could work seven days, if I wanted to, and earn a lot of overtime. For any weeks I worked over forty hours, I got time and a half. A lot of days in the summer, if my crew had trouble with a pour, my supervisor offered me the opportunity to stay on past the end of my shift. I didn't require much sleep, so the extra money was very appealing.

Ted Catanzarite

After my shift was over I used to stop at a bar and have a beer. The drinking age was eighteen, but the bartenders were lax about things back then. They figured that if you were out of high school, then you were old enough to drink. There was a group of us that worked together on the power dam and we used to go out drinking on Friday night. A lot of them were older than me and had come in from the West. Others were in their middle twenties. Many of them were still running around with women. I didn't do too much of that. I was saving my money and interacting with friends from high school and relatives I had in town.

James Romano

The nightlife in Massena revolved around the neighborhood bar. I used to call it the "Bucket of Bud." The guys used to fight in there. It was right by the roadside and it was cold. I would go in there and hug the kerosene stove to keep warm. I didn't dress up. The dirtier I was, the better I looked. The girls that the workers fought over were like something out of a horror magazine. One time a guy asked me to go onto the reservation near Hogansburg. He said, "Come on, we will go and get the chief's daughters!" When we got there, I looked at those Indians with their big heavy coats and big hats on and I didn't make a move because I thought I was going to end up getting killed. I figured if I had my drink and minded my own business, I would be okay. But if I wanted trouble, I was going to get it. If I walked out a door with a light hanging outside there, anybody could have thrown a hatchet at me. So I figured I would stick to the bars in Massena.

I started to go to a bar called the J and L. The owner had dances and a floor show. One night a guy performed there who blindfolded an audience member and shot a card out of his hand with a gun. I volunteered to hold the cards even when a lot of my friends told me not to do it. So I held the card and he put a hole right through it. A few days later, another trucker handed me an article he had pulled from an out-of-town newspaper that stated that the guy had gotten busted in Montreal for shooting off the bottom of a man's finger. When the guy in Montreal held up the card, he put his finger behind it right in the spot where the blindfolded man shot through it. I laughed about it because I apparently held the card right.

I ate a lot of my meals at a diner near Alcoa called The Pierre. Sometimes for lunch the contractors sponsored a cookout at a town park. At night after work some of us would get together there and have our own barbecue. I also liked to go to the roller-skating rink in Norwood when I had the chance. In the winter I would go ice skating in Potsdam because all the college girls went there. I didn't do that very often because I didn't want to drive twenty miles round trip after being in my truck all week. I also enjoyed going to Montreal a few times, but it was a three-hour drive.

Eunice Barkley
When I left the lunch wagon I went to Waddington and worked at the Iroquois Restaurant until eleven at night. When you are young you do a lot of things. That establishment was a hopping place because the owner ran a bar, a restaurant, and hotel. Customers always packed the dining room and bar, including the townspeople who were the regulars. Waddington was such a small town you blinked on the way through or you missed it, so the added population of the Seaway workers made a difference. Six other bar owners also did a great business during the Seaway, particularly Matts, the St. Lawrence Hotel, and Mary Blacks. Most of the time, the

customers got along. There would be a fight here and a fight there, but no violent crime. One of the workers would look at one of the local women and the husband would come in and say, "What are you looking at?" They often settled it outside. Seaway workers were a big boon to the area. None of the locals had made that kind of money before. The workers' paychecks were amazing and they spent them on food and liquor.

Frank Reynolds

It made Ogdensburg a boom town. The workers came in and they had to eat, sleep, and especially drink. Sholette's Bar and Grill on the way into Ogdensburg was always crowded. When I worked at Sparrowhawk Point, it was about nine miles outside of Ogdensburg, so I had to go right past Sholette's every day. I used to stop there for a beer and my wife would stop there on the way home from work. They offered good meals. People came into Ogdensburg and people welcomed them. The locals thought it was great. A lot of them got rooms in hotels, if they got a good rate, or tourist homes. There was also a northside trailer court that opened during the Seaway. A lot of people came with their families. It was quite a lively place.

Ambrose Andre

There were a lot of bars and a lot of drinking because that is what construction people do. I met my wife at a local bar in Massena. She was a teacher and the high school was running two sessions. So she was teaching in the morning and tutoring handicapped kids in the afternoon. She was making more money than I was. One Friday night I went down to Picky's Bar with a friend and there were two gals at the bar. So we chatted with them a little bit and put coins in the jukebox and danced a little. To make a long story short, we picked them both up. The fellow I was with knew them both, so that helped. I took her home and made plans for a date a few nights later and one thing led to another and six months later we were married.

It was really like a summer resort. The St. Lawrence River is very attractive. People fished a lot. I worked six days a week and went to Sacred Heart Church on Sunday. Things like rent skyrocketed. Restaurants opened and then closed when it was over. It was great for that kind of business. The retail businesses, and of course the bars, did well. I used to go to Ottawa and Montreal every once in a while just for the sake of doing something different. I grew up in Lowville, New York, which is a small town, so I was used to a small-town atmosphere. Nobody ever gave me a hard time or abused me. All in all just about everybody was very friendly.

7

The Core of the Corps

SHIRLEY DOXTATER, the society editor for the *Watertown Daily Times*, in an August 27, 1954, article called the wives of Seaway workers in Massena "the core of the Corps." Even though these women never poured concrete or shoveled dirt, each served as a supportive and devoted spouse for their husbands and managed to provide a sense of stability for their children. Like most women in the early 1950s the wives of engineers and equipment operators had married young, had numerous children, and saw mother-hood as the primary source of their identity. Because they had to adapt to living in different locations every few years, these women formed lifelong friendships with each other based on the shared experiences of child rearing that replaced family and ethnic ties. They also joined churches, prayer groups, and recreational leagues, and attended social events as a means of temporarily escaping their husbands and children.

The women who lived in Massena, New York, and Cornwall, Ontario, during the construction of the Seaway struggled to live in a world of new opportunities that they could not take advantage of. For some, this prison of domesticity prevented them from pursuing careers. Many were the first females in their families to receive college degrees and had worked in their professions before their marriage. However, married women who sought employment outside the home faced criticism from mothers-in-law who thought they should be satisfied with being good wives and mothers. Therefore, women bought into the new consumer age and found personal satisfaction in purchasing the best clothes and appliances and in having the latest furniture and decorations.

The Seaway wives claim that the established role for women in the early 1950s would have left them feeling professionally stymied no matter where they resided. Staying at home to care for numerous children while their husbands worked on a project they cherished made them feel lonely and unfulfilled. They also dealt with unfriendly natives at schools and stores and with unfamiliar weather, but managed to make the best of the situation because of the love and respect of their husbands and a close-knit group of female friends. These interviews offer not only a view of the changing life of women during the 1950s, but of the new dynamics of marriage, child rearing, and opportunities in the workplace for females that often clashed with traditional social mores.

Marge Wiles

It was such an experience! When I arrived in Massena in April 1956, there was still snow on the ground. When my family and I had left Kansas it was spring. So that was a shock! My husband, Don, and I knew all about cold weather because we were born and raised in Omaha, Nebraska. However, Massena makes the news at least once a year as being the coldest place in the nation. The first year I lived there it was forty-four below and my husband and I couldn't get over the fact that nothing stopped. Everybody was prepared for it, and they didn't miss a day of work. The weather was a shock even though we were from the Midwest and used to cold weather, but it was very dry and the kids just loved it. Stuff just proceeded.

My husband and I met in high school and got married when I was twenty and he was twenty-one. Don's first job after he graduated from the University of Nebraska was on a dam in a small town in Nebraska with the Bureau of Reclamation. He was on his second job with them in Stockton, Kansas, when he heard about the opportunity on the Seaway project. One of the men whom he worked with on that job left a year before us to go to Massena. We also corresponded with friends we had from Nebraska and learned they had relocated there too. Since we had many friends already there, pretty soon we were on our way. We had helped them move out and when we got to Massena they helped us move in, and this went on for seven more moves. We helped each other out. We stayed friends for all of these years.

I just moved with my husband from job to job and I was always so proud to do it. My husband made $7,400 in 1956, the first year he worked on the Seaway project, which was $1,500 more than his salary from the Bureau of Reclamation. We thought it was a fortune. The contractors had to pay the engineers well because they didn't have any retirement plan. They just had to pay their Social Security tax. So my husband lost out on several years of pension contributions. Most of the men that I knew who worked on the project were our neighbors and they were hard workers. Like my husband, they kept their noses to the grindstone, but they were also very good family men. The one thing that the men always griped about was that they had to work on Saturday morning. They put in a forty-four-hour week, but were home in the evenings so I didn't think anything of it.

Helen Buirgy

I was thirty-three when we moved to Massena in 1955. I had three school-age daughters and during our six-year stay my son was born. I was born and raised in Nebraska on a farm and ranch, and I taught school in a one-room schoolhouse, so my background was pretty tame compared to the adventurous life of moving around to the different projects. Everything, wherever we moved, was an adventure. Every day was exciting and different. I got up every morning wondering what was going to be new that day.

I met my husband, Ralph, at a square dance when I was teaching in Denver, Colorado. He was a native of Denver and was working for the Bureau of Reclamation. He was eleven years older than me and there didn't seem like there was much of a difference in our ages. I think all people are about the same age for about ten or twenty years in midlife. The Bureau of Reclamation hired him when he was in college to work on the plans for the Hoover Dam. He was with the Bureau for twenty-two years and when he left he had been reduced from a GS 12 to a GS 9. He knew that any veteran coming back from the Korean War could have bumped him with or without experience. So he decided it was time to move on, and a job opened up in Massena as an office engineer for Uhl, Hall, and Rich. His work took us all around the country. We had lived in Ohio for ten months and two places in Colorado before we came to Massena. Ralph sometimes worked all night just to make sure the plans were done to his satisfaction. That's what he wanted to do.

Barbara Hampton

I had two children when I moved to Massena; one was ten months old and the other was barely three. I was twenty-seven years old, so I was quite young. I was one of the younger wives. Robert and I felt very fortunate to have two young children in such an atmosphere. I absolutely loved it up there. It was old home week in spite of the fact that I had never met any of the people before. The men and women all had a sense of adventure and a sense of curiosity and wondering about things. The wives were fun to be with. I had no sisters. Somehow the other wives and I became each other's family. Most of us had babies or small children. We established a solid core of support in no time and did things together. I have never had that type of experience since I left that particular group. My family and I were with those people for five years in Massena. That life-loving group of people helped me survive the rigors of life because they taught me that there were some good and bad things about everything. What interested me were all the different kinds of people who worked on that project. Sometimes I would just sit in a chair on my front lawn and watch the world go by.

When we first got up there my little boy and I noticed that there wasn't a child that lived near the river that didn't have a huge crane, a bulldozer, a huge dump truck, and a tractor. When I first saw the lock and dam sites, it was like they were digging up the world with all this incredible machinery. The children had all these toys up there because of all the construction and many of their fathers were on the job. The girls got in on it too. Most families had sandboxes in the backyard with covers on them because of the pets, and the children would build their own version of the Seaway in there.

Joyce Eastin

My husband was the project manager at the Iroquois Dam in Wadding-ton, New York. The St. Lawrence Seaway was the first really huge job he worked on. Initially, after he graduated from college he worked for the Wyoming Highway Department. Then we moved to his native Colorado, where he was employed by a private contractor. After a few years we moved back to Wyoming before a position opened up in Massena. We drove into Massena, New York, on February 22, 1955. We lived in Massena for three and a half years.

On either side of the road leading into Massena, there were all of these intermittent humps. Everything was covered in snow, so we had no idea what these objects were. In the spring when the snow melted, we found out those humps were stone fences. It was really very pretty. Most of the farmhouses were set back from the road, but as you drove by, if you looked closely, you could see clotheslines on the front porches tied to a tree or a pole with a pulley so they would pull the clothes in. It really was very different. It wasn't quite as modern as some of the places I had lived previously. Even though many of the projects my husband worked on were in remote locations, a large city was only a short drive away. I grew up in a really small town in southern Illinois, so it wasn't the small-town life; it was just that Massena was Old Worldy compared to Denver or Casper, Wyoming. It was a culture shock for me.

Several men from the development worked at the Iroquois Dam, so they carpooled. That is why most of the families only needed one car. Most of the workers arrived in Massena with cars because the jobs that they had worked on out West were located in isolated areas. My husband worked mostly eight-hour days, so he had breakfast and supper at home and I packed his lunch. I didn't think it was so bad. He was there to discipline the kids and help get them to bed, which is what I wanted. Howard got along so well with people and he loved the men he worked with, so there were never any serious conflicts that I heard about. He would often talk to me over dinner about what he was doing on the job and how things were progressing. It was steady work and good pay, but working on big construction projects was definitely a way of life. It seemed like we moved every year at that time. But it was actually more like every two years. Howard and I had to pack up the children and move to where the next job site was located.

Irene Bryant

Living in Massena was like trying to exist on the American frontier in 1900. The area was very isolated and residents were very backward.

Many of the children and adults had health problems related to poor nutrition and medical care. There was always a terrible smell that came from the decomposing marine clay; the little kids had bow legs, and they were all short because of their French Canadian ancestry. Many of the women had no teeth and had several children before they were thirty. I think the reason that they all had so many children had to do with the fact that most of the natives were Roman Catholic. One of the incidents I will never forget happened in a local grocery store. I passed a very exhausted-looking young woman in one of the aisles and she had three children and no teeth. I had never seen an individual like this before and I was just horrified. The priest from one of the local Catholic churches came up to her and asked her how she was. I am sure he had had a good night's sleep and he was just so jolly and happy. The woman brightened up immediately. That made an impression on me! There also must have been a deficiency in minerals or a contaminant in the water. I had a friend who was a nurse at St. Vincent's who said that there was a high percentage of children born with mental retardation or some other type of birth defect. I guess she didn't know what caused it. But when I drove around I just had the feeling that the town was terribly polluted. There was something terribly wrong.

The weather was another issue. It was hot in the summer, rainy in the spring, and bitterly cold in the winter. All of these extremes caused problems for my husband and me. When we arrived in Massena, the builder told us we could only move our belongings into our house at night or very early in the morning before the ground thawed because the driveway was muddy. The moving company driver could not pull his van up to the garage unless the ground was frozen or else it would get stuck. Then it was forty below one day when Jack was in New York City on business. The way the houses were built back then, the oil heater was in the garage. The temperature was so cold that it was causing the oil to sludge, which could have left us without heat. My neighbors from across the street were worried to death that my children and I would freeze. The husband was nice enough to come over and install a light to heat the oil to stop the sludging. I had many unforgettable experiences in Massena!

Ann Marmo

My family and I lived in Massena for eight years from August 1953 to October 1961. It was the longest eight years of my life! Joe and I moved to Massena from Colorado. Colorado was a nice place to be and Massena was so different. The Colorado weather was so much better. It was a real change. We moved there in August when it was hot and humid. I was used to nice dry, cool summers. Basically I wasn't a happy camper in Massena. The one thing that I used to say frequently was "we are ninety miles from everywhere," because we were ninety miles from Watertown, New York, and ninety miles from Montreal. The hospitals and doctors in Montreal offered the best services at that time. If my husband or family members or our neighbors were terribly ill, we traveled to Montreal or Watertown for hospital admissions. It just seemed like we were isolated. It was kind of a culture shock.

The residents' social outlook and the local economy were behind the times. I feel that was probably typical of many small towns during the postwar years, including my home town of Hornell, New York. A lot of my high school classmates married one another and still live there. A lot of my female friends became housewives right after graduation and didn't continue their education. My goal right from the start was to get out of Hornell and become something. A lot of these women in isolated areas didn't seem to have any career aspirations and I don't think they still do. The women I met in Massena were the same way, so maybe it was not an unusual thing.

Looking back now, I realize that I would not have been contented anywhere during the raising of my little ones. Maybe if I had not had three children, it wouldn't have been so tough. I had a three-month-old daughter and a four-year-old son when I arrived in Massena, and then eventually I had another baby. I felt so confined during those horribly long winters. I felt like I was just existing and I used to say, "I feel like I am living in Siberia," because the snow would pile up and in order to get the car out of the garage I would have to go out and shovel before I could get the three kids in. It was just a hassle. The severity of the weather was something I had never experienced before. I knew what winter was, but it was so dark, cold, and depressing in Massena.

Phyllis Cotter

I was twenty-seven and Jim was thirty when we moved to Massena. My mother thought I was moving to the end of the world and she was never going to see me again. My daughter was six weeks old when I got there and my son was a year and a half old. Then I had another son about fifteen months later. I was pretty much stuck at home with three little kids. Jim and I were there from 1955 to September 1958. I remember the four-day drive to get there. I had two babies in diapers and I was hanging over the backseat of the car changing somebody's diaper about half the time. I had a basket in the backseat for the baby and the youngest was just on a pad with no seat belt. My husband built a platform for them to sleep on between the two seats. It was a journey I will never forget.

28. Jim and Phyllis Cotter. Courtesy of Phyllis Cotter.

Shirley Davis

My husband, Dick, and I were originally from Maine. We moved to Waddington, New York, in 1952 when Dick got a job at Alcoa. At the time there was no housing for us in Massena, so we found a house on the river several miles away. We didn't move into the housing that Uhl, Hall, and Rich built for the engineers in 1955 because we liked where we were living. It was so beautiful in Waddington. Our house was on the water and we could watch the ships go by. It was very exciting living there because we were right on the banks of the St. Lawrence River where they were doing a lot of the construction on the Seaway. So we got to see it.

As an artist I found a lot of things to paint. I didn't create paintings to sell at that time; I did it mostly for my own benefit and for the benefit of those around me. I read books all winter and painted and I had a garden all summer. I really enjoyed the life there.

Dick had graduated from the University of Maine with a degree in mechanical engineering, so he was excited about the Seaway project starting. He heard about a job with Uhl, Hall, and Rich through friends he knew. My husband really liked his job. I think there was a lot of goodwill among the employees of the different engineering companies. A lot of them came from the Bureau of Reclamation out West and knew each other from their jobs on the Hungry Horse and Libby Dam projects. Other engineers came from the South. I don't know if many of them were used to the weather, even though there are some pretty cold places in the West. On cold mornings, Dick said there was steam rising off the river. The weather didn't bother me. Sometimes, when I had to walk to the grocery store, it might be twenty below, but it was right in the neighborhood. Also, being from Maine, I was used to the fluctuating weather.

Dick and I had close friends who lived in Waddington before we ever moved there, which made it easier. I also socialized with some of the other engineers' wives, like Barbara Hampton, but mostly local people. We had nice next-door neighbors who were very friendly and were transplants like us, being from Tennessee. Another couple down the street had moved to the area from Colorado. Having supportive people around was important to us as we had no relatives that lived close by. I also went to the Presbyterian church and was very accepted by the other church members and was very active. However, playing bridge was my biggest social activity. Waddington and Massena were both great bridge-playing towns. It was a very popular game back then and was a very social game.

Many women from Cornwall and Massena married the transient workers who temporarily lived in the area to work on the Seaway and Power Project. The number of eligible bachelors in the area had been declining with the loss of jobs owing to the recent downsizing of the towns' textile and aluminum industries. Some of the workers' southern accents fascinated many of the local women, who had never heard them before. Unfortunately, the fascination wore off after the women had experienced the workers' migrant lifestyles for a few years. They didn't know how difficult it would be to be away from their families and to move from one town to another every few years. Some of the women, however, embraced this new adventurous way of life and never returned to their hometowns.

Melba Singleton

I was the chief operator on the second floor of Bell Telephone in Cornwall, Ontario. My friends and I would go over to Massena for the evening to meet the workers and dance with them. There was a young man called Duane Johnson. We called him Buddy Buddy Johnson and I often danced with him because his parents were personal friends of mine. My best friend, Bessie, had a little black car and five of us would drive over the border to have a good time for a few hours and then go home. We were all in our late twenties and looking for husbands. I was still living at home. Bessie played the piano, so we went to a bar where there was a piano and got Bessie to play. We just had good all-around fun. Cornwall didn't have drinking establishments where both men and women could go and dance and have a drink. I never drank alcohol, but my girlfriends got men to buy them drinks.

I met my husband, Ray, a heavy equipment operator, through my girlfriend. We started going together in December 1955 and got married in November 1956. I was twenty-nine at the time and Ray was twenty-eight. Ray was on a date with my friend and I was with his oiler. That night Ray and I sat together and talked most of the evening. After going on three dates with my friend, Ray broke up with her and called me. Ray was from Conway, South Carolina. I always loved a man with a southern accent. There had been a lot of southerners working on a pipeline in Canada and I would get them on the line and just listen to them talk. I was taken with Ray when I met him because he was a very quiet man and he didn't drink.

Ray's family members all lived in Massena during the Seaway project. Ray, his uncle, father, and two brothers were equipment operators and had worked on the construction of numerous dams and waterways. Prior to the start of the building of the Seaway, they had all operated machinery on a road project in Ohio. When that job came to an end and they heard that the contractors needed operators, all of the family members packed up their belongings and relocated to Massena. It was a family concern. Each family member and his wife and kids had their own trailer. In Massena they all rented plots in the same lot and parked them next to each other. Other men operated heavy equipment, but Ray, his father, and his brothers were well trained and were all excellent operators. His father trained them all, including Ray, to run a dragline in the Florida swamps. He started off as an oiler on the machine and then his father taught him how to operate each piece of equipment. He taught all four boys to run those big machines, including the crane. Ray worked on the Long Sault until it was done and then on the big power dam. Other workers envied these five men because of their skills and the respect they had from the contractors.

After I got married I quit my job at the phone company and stayed home. Ray's family opened a restaurant for eight months. The first mistake I made in my marriage was going into business with my in-laws. I had a mother-in-law that topped all mothers-in-law. She was not a nice person. My father-in-law was the opposite; he was an angel. She was fourteen and he was eighteen when they got married. Ray's family decided to open a restaurant in a building owned by the operator of the trailer park. They wanted to make food for the large number of southern workers on the project, so we made all of the traditional dishes, including dressing and grits and eggs. In the morning I would go and prepare lunch with Ray's aunt and then I would go back to help with the dinner preparation and work until closing. That was a lot of hours and it was really hard work.

Many of the engineers who worked for PASNY's main contractor—Uhl, Hall, and Rich—lived in the Buckeye Development. The contractor funded the construction of these small homes with both public and private funds and rented them to company employees for a small fee. Workers and their wives filled these typical ranch-style homes with appliances that were part of the new American Dream. Initially the Corps of Engineers had planned on constructing housing for all the workers. However, the administrators of PASNY and the Seaway Development Corporation, specifically Robert Moses, did not want to waste construction funds on temporary dormitories. Moses felt that the towns of Massena and Cornwall and those in the surrounding area would be able to absorb the added population, and if not, town officials should shoulder the burden of building additional accommodations.' Therefore, most of the transient laborers and carpenters and their families set up trailers in one of the many parks established by local farmers. The remaining workers rented homes, rooms, or apartments in Massena or the nearby towns of Waddington, Canton, Madrid, and Potsdam, New York.

Helen Buirgy

My husband and I were one of the first couples to move into the Buckeye Development in January 1956. At that time, there were no paved roads or driveways. The only way to get to our house was to drive through muddy paths covered with ice. The moving van we hired to transport our belongings to Massena broke through the ice and got stuck. The driver and his helpers could not unload our furniture until they figured out how to get it out of the back of the truck and carry it the rest of the way to our house. The prefab homes Uhl, Hall, and Rich built had wood paneling and only three bedrooms. My husband and I and our girls always enjoyed moving and we were happy to have a new home, even if it was small.

Phyllis Cotter

When my family and I first arrived in Massena, the engineers' houses weren't done, so we lived in a little apartment above a garage with no air conditioning for six weeks. It was really hot and there was no washing machine, so I had to wash the baby's diapers and our clothes in the bathtub and hang them over bushes to dry. When we finally moved into the Buckeye Development, we didn't have electricity in our house. So Jim extended a cord to the next house so we could have lights and run our appliances. In the middle of the night, I had to fire up a camp stove to warm up the baby's formula. After a few months, our electricity was hooked up.

We lived in the first house on Middlebury Street behind the creamery. I could walk over and get the milk I needed, as well as ice cream. On previous jobs my husband and I had lived in established neighborhoods. So not only was our house old, we didn't have neighbors who had relocated to work on the construction project. Instead, we would have to drive to see our friends in a town that was fifteen miles away. So for me, it was nice to have neighbors in Massena who were all new to the area and whose houses I could walk to easily for a cup of coffee. Buckeye was so great because there were so many of us from the West and we all stuck together.

Joyce Eastin

The Buckeye Development was a community within a community. All of the families that lived there were foreigners, imports. Our husbands all worked on the job and because we were all one-car families we did a lot together. There were a lot of children my children's age. I had three children when I got there and none of them were in school, so that is why I stayed at home a lot. My oldest one went to school that fall. We moved in February and he went to kindergarten in September. Since the homes were close together, the wives would get together. We would visit, have lunch, and watch the kids or go out to lunch or shopping now and then. When my kids were at school, I could clean my house in no time because it was so small and still have time to attend a coffee klatch. Sometimes just two or three of us would get together and have a chat. Then it would be time to go back home and fix lunch for your child if he or she was coming home from half-day kindergarten.

Marge Wiles

My husband and I thought that living in Buckeye was a lovely experience. The homes had three bedrooms and one bath, and I thought I had moved into a palace. Our ranch was only one thousand square feet, but we were just tickled pink with it. We rented the house for an extremely low amount. There were about three rows of twenty houses, so there were seventy families. When we arrived, our house was ready for us. It was really neat. The builder had sent me the window measurements, so I stopped in Omaha where our families were and went down to one of the department stores and ordered my drapes. By the time I got to Massena, my drapes had arrived.

We lived across from the Cotters and their third child was born there. I watched their two other children while Phyllis was in the hospital. I was

29. Coffee party for wives of Uhl, Hall, and Rich engineers at the Buckeye Development. Courtesy of Kenneth Hallock.

on the phone talking to her when my fourth one was due and my water broke. So we just have a lot of funny memories of our time there. The Cotters lived across the street from us in one direction and the Potters lived behind me in the other direction. We would get the names mixed up sometimes when we were talking. We were all raising children. The kids just kept coming and each one was going to be our last.

Barbara Hampton

In the housing that Uhl, Hall, and Rich constructed, the windows had a metal sill. Nothing could be worse than a metal sill on a window because it transmitted the cold. The windows sills would freeze over like the coils in the old refrigerators that had to be defrosted. I made the mistake of hanging up my good draperies, and no matter how much I turned up the heat in the house, this layer of ice formed. Then when the sun hit it, of course the ice melted. My draperies absorbed some of the water and became discolored. The house didn't even have painted walls. They were covered with just plain finished boards—not nice paneling. There was no way to paint the walls another color as a way to cheer myself up in the winter. I didn't live in Massena very long before I realized that the weather had a personality of its own and I had to learn to deal with it.

Ann Marmo

My husband and I never lived in the housing projects on any of the large construction sites he worked on. Joe and I realized after we left Massena that we preferred living in our own home and not in the contractor-sponsored housing because we wanted to be more independent. We love people and we love groups, but realistically we liked to have our house away from the mainstream. The houses in Buckeye were small, so we were better off. I didn't like our house because it was so hard to heat. Our oil bill was more than our rent. We paid $75 a month rent for a furnished house. I remember one of those extremely cold winters our oil bill was $80 and the oil was $0.15 a gallon.

Our power went out for a week during an ice storm in the winter of 1956 when it was forty below zero. Our house had these big rooms, including a huge kitchen. My brothers were living with me at the time and used these old army quilts that the owners had left in the attic to cover the doorways and keep the heat in. We also put them on the floor to sleep on. The kitchen was equipped with a side-arm stove where we burned coal and wood to keep warm. Then luckily our power was restored. Unfortunately one of our neighbors was still out of power, so they moved in with us with their baby for five days until their power came on. That was one of the worst experiences I have ever had.

Joe and I were very kind to my brothers. They were great guys. One of them had just gotten out of high school and we had him come and live with us for two years and when my other brother got out of the service a short time later, he came to live with us too. So I had my two brothers, my three kids, and of course Joe, and I was working myself too hard. That was my own fault and had nothing to do with Massena itself. That was just the way I was living and it was not good for me. My brothers were both making more than Joe and living with us. When they both came to live with us, I charged them for their food because when I figured out how much it cost to feed them twice a day and provide them with a lunch, it turned out it added up to $150 a month. If that had been today, I think I would have been smart and had them make their own lunches, but you see back then women were expected to do everything. It was just awful! They lived with us for a couple of years, but at one point I just got too exhausted, and we decided to ask them to move out. I just said to Joe, "I don't think I can keep doing this!" So they rented a room nearby. That is when I found out I was having my third child. So I have to say my life was never boring because I was always busy.

The creation of trailer parks seemed to be the best solution to the housing shortage in Massena because most of the workers did not plan to stay in the area permanently. If that had been the case, then housing developments would have been the answer. Although many workers lived in trailer parks, some managed to rent private lots in the country to set up their trailers on. Some of the wives were not fond of the trailer parks, but there wasn't anything else suitable in Massena. Others worried about exposing their children to alternative and what were viewed as immoral lifestyles. They didn't want their children to see and hear what sometimes went on in group housing where a lot of drunk and vulgar men shacked up with women who weren't their wives. However, most workers felt it was the only way to live in the construction business, where costs were high and housing was scarce.

Melba Singleton

Ray and I were trailer trash. We bought a brand new trailer when we got married. There were local people who built trailer parks and we were all one big happy family. Everybody got to know everybody, and I never had any problems with people in trailers. There were people from Virginia, Maryland, North Carolina, South Carolina, Tennessee, and Maryland. Of course, people lived in the trailer park whom I was not too fond of, but I didn't have nothing to do with them because I am a firm believer in live and let live. There was drinking and all that stuff going on and I didn't associate with it. A lot of the construction people there drank, but we stayed by ourselves. I can remember reading in the paper about them expecting the worst from the trailer trash.

Our trailer was insulated, but one morning I woke up and we had no heat. It was fifty-six below zero. Ray got out there and took the heater all apart. Luckily we had a second propane tank that we could use and we had the heat on in no time and warmed up, but gracious it was terrible. Everybody else's trailers froze up.

Many of the wives on the project socialized with each other as did their children. Because of their shared circumstances and closeness in age, they formed a make- shift support group that played cards and shopped together and provided child care in a foreign and isolated atmosphere. Their husbands worked long hours, so if emergencies arose the Seaway wives relied on the kindness of their neighbors. Many of the sons and daughters of the Seaway workers had been born in the West and appreciated the opportunity to experience the joys of winter. Also, owing to the longer duration of the project construction, they made lifelong friends with their neighbors. As was typical of women in the 1950s, these women had to adapt their lives to their families' needs, and they illustrated an enormous commitment to the well-being and happiness of their children.

Phyllis Cotter

The other wives and I organized a bridge club. It gave us all something to look forward to doing independent of our husbands and kids. All our kids also played together. Behind all of the houses there was grass, so our kids could go out in back and play in a big open area. Our neighbors had kids the same age as ours and even when they were young, I allowed them to walk to their friends' houses because I knew all the people in the neighborhood. They didn't have to have their mama take them down to their friend's house to play; they could walk there by themselves. That was very unusual for kids that young to have that kind of freedom.

Marge Wiles

I kept getting pregnant and my husband and I continually said that each one was going to be our last. My children at the time were eleven, nine, and four. My youngest was just ready to start kindergarten. Then I had a little girl while I was there . . . you know those cold winter nights. To avoid loneliness, I joined in a bridge club with two tables. My husband stayed at home with the kids at night. It was a break for me to get away from the kids. It worked out real well. Six out of eight of us had husbands who worked for Uhl, Hall, and Rich, but there were also some wives of men who worked for the other contractors. Every single one of us had a tail-ender. That was what Massena did to us.

My kids loved it there. They loved the winter sports and the snow. We didn't have any fences, so there was an ice skating rink that was flooded in all the backyards that continued to the other end of the street. We all knew each others' kids. I just thought it was a wonderful experience. My kids made some lifelong friends. My husband and I didn't think it was hard on our kids at the time, but my oldest daughter said she would never do that to her kids. She hated leaving the friends she made who didn't come with us. She made some good friends in Massena, and they left a year earlier to go to the Niagara Falls project. My son's high school graduating class in Orville, California, included seven kids who had started kindergarten with him in Massena. About twenty-seven families from Uhl, Hall, and Rich followed us from Massena, to Niagara Falls, to Orville, California, so the parties continued. It was just great.

Barbara Hampton

Well, it was like one big family. Helen Buirgy was older than me, and her daughter, Bonnie, was eleven years old, and she always wanted to be a schoolteacher like her mother. Because we had so many small children, we had to make sure they had something to do every minute all summer. So Helen set Bonnie up in their garage with little chairs and a big blackboard. The other mothers and I would bring over our little children, and they would sit in their chairs for about an hour. Bonnie Buirgy would read to them. Then on some afternoons when I took care of a lot of the kids at my house and they were getting restless, I used to take them outside and read Rudyard Kipling's *Just So Stories* to them on the lawn. It was just wonderful. To this day one of the gals told me her son remembers hearing me read stories to them. We went on picnics and played games, and one time we enacted *The Wizard of Oz*. We just had a marvelous time, and I enjoyed my children. I think you have to make the best of where you are at the time.

Joyce Eastin

My kids, all three of them, loved to play outside, especially in the snow. The first winter we lived there, the snow was so deep that the kids built tunnels in the front yard. They would crawl through them and have a ball. We also had the neighborhood ice skating rink which the fire department flooded between the two rows of houses from one end of the street to the other, and the kids would skate out there all the time. Of course we all had barrels in our backyard where we burned our trash, and we would light fires in them so the kids could go and get warm every once in a while. In the summer the kids decorated their bikes and dressed up and had their own parade up and down the street. My kids just thrived.

Churches were the main places where townspeople and the construction families interacted. The arrival of Seaway workers in Massena in 1954 led to significant growth in the membership of local churches and introduced new faiths to area residents. The leaders of most denominations recorded increased attendance at church services and Sunday school classes. Many local parish councils financed long-planned expansion projects with the generous donations of Seaway workers, and reverends and priests constructed new chapels near worker housing and added extra services to coincide with Seaway workers' schedules. Most of the wives said that they felt welcomed by the parishioners. Couples agreed that they wanted to set a good example for their children and provide lifelong social and religious values.

While the Catholic Church remained the area's largest congregations, many supervisors, contractors, and their families attended Presbyterian and Episcopal services. The influx of Seaway workers from the southern and western United States enlarged the membership in more evangelical faiths, including the Baptist and Congregational parishes, and also encouraged ministers from several unrepresented religions, including the Assembly of God and the Church of Christ, to come to Massena to establish congregations. For the workers who constructed the St. Lawrence Seaway and Power Project and their families, religion remained the one constant element of their lives. Once families or individuals located a place to live, the next thing they did was join a parish. Workers and their wives also became active members of the choir, women's auxiliary, and vestry.

Marge Wiles

All of the project families became members of the local churches. This was the same at all of the projects we moved to. My husband became an elder in the Methodist church and I was very active in the women's circle. The parishioners were very friendly and welcoming and made us feel like we had lived there forever. Being an active member of a church was something I had done all of my life. I have always been involved in church activities wherever I have gone because my mom always was. I loved to tag along when she cleaned the church or went to the women's dinners. I thought it was really neat to help out. I was just raised that way. It is just something I felt I had to do. Don and I just loved our minister and the people in the church. We both thought that it was important for our children to be part of one stable institution even though we moved a lot.

Joyce Eastin

My husband and I attended the Baptist church on the corner of Orvis Street. It was one hundred years old when we were there. Howard and I had always gone to church, and many of the families that moved from one project to another wanted to keep their faith one of the constants in their lives. Those of us with children wanted to provide them with a strong religious foundation that they could pass on to the next generation in the same way that our parents had done for us. With very few other stable institutions or organizations in our lives, churches seemed to be the one thing that I could find near each project site. Most of the workers and their families attended the same churches that they were born into. If they went to church where they came from, they went to church at each project site. My kids also attended the Baptist Bible School in the summertime, which was about the only extra activity offered by our pastor.

Helen Buirgy

On other projects that my husband worked on, the workers and their families attended the Baptist or Methodist church. I was raised a Baptist. I think we all worship the same God. However, wherever my husband and I moved, we visited different churches to see which one we liked the best. Once we settled on a parish, I always taught Sunday school and my husband sang in the choir. He had a trained tenor voice. When I was living in Massena, my main social event of the week was going to church because the rest of the week I was busy taking care of my children. Bob and I first attended the Baptist church, but I didn't feel comfortable there because I never felt that I was dressed well enough. In Massena, many of the workers and their families went to the Congregational church, so we attended a few of their services. After we drifted around for a while, we initially chose the Presbyterian church. That all changed when my little girl got hit by a car. The ladies auxiliary from the Methodist church came over and sat with me and provided me with support until my daughter recovered. These women became close friends and I have been a Methodist ever since.

What surprised me in Massena was that 80 percent of the residents were Catholic. I was not familiar with the traditions and practices of the Catholic faith. One day my husband came home and he was laughing. He had breakfast with some of the bosses, and he almost told one of them that he had dirt on his face. He was happy that he had kept his mouth shut because when he got back to the job site one of his co-workers who was Catholic explained that the man had been to a church service where they put ashes on his forehead.

Ann Marmo

Joe and I went to the Catholic church. In those days it was not customary for the parishioners to do a lot of the work in the church, because they had so many priests and nuns, so we just went to church. It is not to say that we couldn't have participated, but that was not the customary thing to do. Our son attended Catholic school and at that time it was supposed to be the ultimate in education.

Barbara Hampton

Bob and I went to the Congregationalist church because we found it the most interesting and the parishioners the most welcoming. I not only enjoyed the service and the pastor, but also taking all of the classes that they sponsored to study comparative religions. Some of the professors came from St. Lawrence University for what they called a university of light. It was so interesting and it helped with the winter weather. I looked forward to not only attending the lectures, but also reading the weekly assignments. It gave me something interesting to do.

The relationship between local residents and the construction workers and their families was another dimension of life in Massena from 1954 to 1958. Before the commencement of Seaway construction, the local police agencies of Ogdensburg and Massena anticipated a rise in the murder and crime rate. Local newspapers published editorials whose writers bemoaned the impending invasion of their small town by interlopers and rednecks. Throughout the project, the workers' wives had to deal with suspicious and unfriendly shopkeepers, clerks, and even druggists. As business owners and local residents treated these women like second-class citizens, they had as little interaction with the native population as possible. Many did not understand this harsh treatment, as they spent money at local stores and supported local churches and voluntary organizations. One wife stated that she should be as much a part of the community as possible even in the face of resentment.

Ann Marmo

When Joe and I first arrived, the editorial page was filled with letters whose authors complained that all of the workers and their families coming to the area were riff-raff and were going to ruin their nice little community. I guess they estimated with all the construction workers on all the Seaway sites, about ten thousand people were going to settle in Massena, Cornwall, and the surrounding towns. Realistically, it was a fantastic group of people who improved the area a lot through their volunteer work, donations of money, and spending of cash on normal expenses. At first it was hard to see the editorials, but eventually Joe and I said, "These people don't know what they are talking about," and we let it go because we knew that we were all good people. The thing is, typically, we like to believe professionals are good people, but contractors also brought in a huge amount of laborers. There is nothing wrong with laborers, but sometimes they are a lower class and that might have been a problem. But I honestly don't think the police or the locals ever had too many problems with the workers. My husband and I felt after we lived in Massena for a few years, all of the fears and suspicions subsided.

Shirley Davis

I remember there was quite a bit of both positive and negative publicity in the local paper about the project before it started. I sat on the zoning board in Waddington, New York, hoping to make plans for the possible influx of people. For a small town, the impact of a few hundred people was a concern. Members of the board thought there would be a lot of development and they wanted to control the location and type of residential and commercial buildings. They wanted to appropriately position filling stations and trailer parks so that they would not disrupt current residential areas. There was a fear that the town was going to change in a negative way. People were suspicious of strangers. I don't remember there was a lot of animosity or anything like that. Most of the people who moved in came a few at a time and most of them seemed to settle in pretty well because a lot of them were engineers or professionals. It was a change, but I think people took it in stride.

Barbara Hampton

I heard later on from the townspeople that they were all afraid that we would be like people from the Wild West, and I was a bit surprised at that. Local residents became more friendly when they found out that the workers and their wives went to church and volunteered for the PTA and the Little League. They weren't unhappy with us at all when we all got there and they found out that we all had one nose, two eyes, and two ears.

Mr. Levine owned all the stores back then, and I thought he was a nice man. He was funny and he gave me his fix on the matter. As far as he was concerned, all of the new residents and tourists increased his business and he had made a lot of money. He also was able to stock a better quality of clothing. I was just glad that he liked us being there. I never had any idea that the general population was afraid of us because the members of our church were so welcoming.

Marge Wiles

I don't think it was all peaches and cream. I think some of the natives resented us. Of course my husband and I were used to it because of our experiences with the Bureau of Reclamation. You have to like people and you have to enjoy making new friends and we both did. Pat O'Brien was a townsperson and she was my neighbor and we became good friends. Her backyard backed into my side yard. I think you can learn a lot about your neighbor by what she hangs on the clothesline. Our clotheslines were attached to the telephone poles and they went straight out over the snow. When I was hanging out all those diapers, Pat said they were twenty-four badges of surrender. In our case, I don't remember any problems with the local people, or with any of our neighbors.

The workers and their families brought a lot of money into that town. I would fill up one grocery cart until I couldn't put any more in it, and then I would park it and fill up another one. I remember how excited I was when they built Loblaws. My husband at that time had a really big appetite. But my neighbor, Joyce Eastin, was quoted in the local paper complaining about the lack of selection in the stores. The article gave her name and I was just aghast at her snobbish comments. She was very critical of the shopping. I didn't think it was that bad.

Joyce Eastin

None of us ladies who lived in the Buckeye Development really thought that they (the locals) wanted us to be there. The clerks waited on us in the stores, but something happened that I will never forget. Howard, my husband, got sick. So he went to the doctor and was given two prescriptions. When he took them to the drugstore on a Sunday afternoon, the druggist filled one of them and then said that since it was time to close, he would not fill the other one. So Howard had to wait until the next day to get the other one filled. We thought it was appalling since Howard's dad was a druggist, and he knew you didn't do things like that. But they did in Massena!

Another strange thing happened in Levine's, the local clothing store. Late one afternoon one of my neighbors was shopping there and had picked out several items to buy. She had been paying so much attention to finding several dresses that she had lost track of time. When it got close to closing time, the woman managing the store said, "I am sorry. You will have to come back tomorrow." She shut her cash register up and that was it. They made us feel like outsiders, but that didn't stop me from doing what I had to do.

One time I had to get all my Christmas shopping done and get it mailed by December the fifth to get the gifts to all our relatives in Colorado. So I got a babysitter and went to that same department store. The mothers all exchanged taking care of each other's kids while we did chores or ran errands. So I dropped off my kids who weren't in school with someone and went shopping. When I pulled into the parking lot, there were only a few cars there. I went in and said, "Hello" to the clerk and she said, "What are you doing out? Don't you know it is twenty degrees out there?" To tell you the truth, I didn't realize it. I did what I had to do, and that was all there was to it.

Some of the locals at the church we joined were very friendly and very nice, but if I went into one of the local stores or restaurants, I didn't find the same friendliness. There were a lot of Massena residents who had emigrated from Eastern Europe and were not very receptive to change. They wanted their town and their lives to stay the same.

Melba Singleton

My main contact with the native population of Massena was with the customers at our restaurant. The only thing I couldn't stand was the drunks. We would stay open until two in the morning and they would come from all the bars in Massena to have breakfast. I was very strict and I tried to be nice if they got unruly. I would try and talk to them and tell them we didn't allow that sort of behavior. A lot of them got up and left, and that was okay. But then we had others that stayed and finished their meals. I tried to be discreet about it and talk to them until they calmed down, because I sure did not want any fighting. One night we had some of the Indians come in from the reservation and they got fighting among themselves and it was terrible.

Living in Massena exposed Seaway workers and their families to new leisure activities and altered men's relationships with their wives and children. The married workers and their families enjoyed the various recreational activities in Massena and in the nearby Adirondack Mountains, including hiking, camping, fishing, and sightseeing. Many men and women had fished some of the most famous streams in the Rockies and the High Sierras, and had skied the slopes of Colorado, but they relished all that the area lakes and mountains had to offer. The location of workers' housing in Massena, which was close to the project sites, also offered a unique experience for workers. Before relocating to Massena, many engineers and carpenters had not been able to bring their families with them to various public works projects because of the lack of housing, or else they had had to drive a long distance to work, which limited their leisure time. Having a shorter commute offered men not only the opportunity to spend more time with their children, but also to relieve their wives of their childcare duties at night and on the weekends. Many men took a more active role in the upbringing of their children, and the extra time spent with their spouses strengthened their marriages.

Marge Wiles

Even though Don made a good salary, my family and I spent a lot of time camping and taking day trips in our car and exploring the area because that was about all we could afford. My kids have wonderful memories of our camping. On one occasion, I suddenly realized that we were not the only family taking advantages of many of the opportunities the area had to offer. One weekend several of us from Buckeye went camping in an isolated area and we almost couldn't find a spot to pitch our tent because it was so crowded.

I also got to share the beautiful surroundings with many of our relatives who visited us while we were there. My mother came from Omaha and spent the summer, and she was just entranced by what she saw because it was so different from Nebraska—the scenery, the attitude, everything. She had been a widow for many years and after we left she didn't do much traveling, so she welcomed the opportunity to come to Massena and visit us and the kids. I took her shopping and to some of the historic places and she just thought she was the luckiest mom in the world to see the natural beauty as well as the big construction sites. Whenever we had relatives visit, we were just so proud to take them out and show them all the construction. Don and I were also grateful that he actually had some vacation time, which he had never had on other projects. It allowed us to travel to Canada to see the Parliament buildings and other sites along the way. I thought it was wonderfully educational for the children.

Helen Buirgy

My husband and I loved living in Massena! We both felt that the area was the closest in terms of weather and scenery to Colorado, which is where we had decided to retire. We both liked the cold and the winter and the fishing and beauty of the St. Lawrence River. Ralph and I had a wonderful relationship. He was a great father and loved doing things with me and the kids. The kids were never healthier than they were in Massena. They would come home from school and change into play clothes and go out and climb and slide in the snow and come in with pink cheeks.

Ralph and I did a lot of outdoor things with the children. Every weekend we either went fishing or hiking or found something new to do outdoors. Ralph often found new places to fish when he was at work. He never took us all fishing unless he had investigated the site first to see that it was safe. I even remember taking some of the kids out in a boat when they were too young to swim, but we were all having so much fun, I don't think it concerned us. I always wondered when he got all of the weekend activities planned. Recently I began to think that it wasn't his idea at all. It was probably mine. I would have everything done around the house so we could go outdoors, and I may have been the one who thought of the many activities that we eventually did. Everywhere Ralph and I moved, it seemed like a vacation. We both tried to expose our children to new things and to make their lives fun.

Barbara Hampton

Like any little town, there was more to do in Massena than first met the eye. I think many of the women I lived near did not take the time to look for new activities to pursue. I continued to read books, which I had always done; however, after I lived there for a while, I realized how much there was to do. There is always something to do in a small town. I just had to look under the covers of the houses and churches. Since I was near a college, I discovered that there were plays, art exhibits, and other activities that were open to the public. I also enjoyed ice skating with my children and there were numerous rinks and frozen ponds. My kids and I used to also have a lot of fun skiing at the nearby slopes.

With my husband going out on the water almost every day, he found all the inlets and tributaries. He found one near Waddington where the ice would freeze extremely thick in the wintertime. I recall going out at night with a few other couples and building a bonfire on that ice. It was that thick. It was so cold that when we set the pickle jar down while we were roasting hot dogs and marshmallows, it froze. I learned to live with the cold.

I had just as much fun in the summer as I did in the winter months. In the summer there were lots of facilities if you were campers, and we were hikers, canoers, and backpackers before it was fashionable. For our honeymoon, my husband took me on a canoe trip on Long Lake in uncharted waters in an area where there was no one around. He had packed all our provisions in the canoe and we encountered another young couple doing the same thing. I thought, "Thank God I was a Girl Scout!" He never had to teach me how to canoe or hike, so all I can say is I spent my life keeping up with him. We had a wonderful time.

Phyllis Cotter

Jim and I really enjoyed Massena and all that the surrounding area had to offer because being from the West we were used to driving for miles to see anything. Up there we could drive just a short distance and see all kinds of interesting things. We made a trip to Montreal and to Quebec and to Fort Ticonderoga and to Upper Jay—the make-believe land—and over to Plattsburgh. We loved the covered bridges and the antique stores out in the country. The weekend trips were so great. My most memorable trip was when we drove through the Adirondacks to see the fall foliage. I grew up in Montana, so the only trees I was familiar with were pine trees or evergreens, so seeing the leaves turn in the fall was so unusual for me. I can still close my eyes and see all the wonderful colors I saw on our drive. Jim and I also got to drive to Quebec City by ourselves for the weekend because one of our neighbors offered to watch our kids.

Irene Bryant

There wasn't much for women there. We had our bridge games, but there weren't any good shopping centers or cultural events. There was a professional women's organization and the locals were kind of nasty. I met a woman who had graduated from Cornell and her husband managed a milk-producing plant. I mentioned her name to one of the officers of the club as a potential member and she said we couldn't have anyone who lived out in the boondocks as a member. The woman was probably better educated than she was and a lovely person. I also took a French course and the first six weeks concerned the Roman Catholic Church, learning the various practices and beliefs, so that was the end of that. I am not religious. I remember traveling up to Montreal; we would go through these dirt-poor towns with a beautiful church. I guess worshipping in such an ornate place calmed the parishioners and gave them hope and the will to live even in those destitute conditions.

Ann Marmo

One of my favorite places to shop and get my hair done was across the river in Cornwall. I don't think I thought there were any good beauticians at the time in Massena. In those days the prices were much better in Canada than in the U.S., especially on woolen goods and clothing, so some of us would go over for the whole day and have some fun.

We were different than the local women. Levine's was the only clothing store in Massena. What clothes we didn't buy there we ordered from catalogs. I practically lived out of a mail-order catalog, either Sears or Wards. Another interesting thing was that in our neighborhood we had a store which delivered groceries for free. I very seldom went shopping because in those days they didn't have shopping carts where you could put a child in it and push them around. So I would just order my groceries once a week. Most of us couldn't afford much. Joe was making $5,600 a year when we first moved there and by the time we left, he was bringing home about $8,000. The construction people made more than the engineers and the professionals.

I started to sew when I lived in Massena. Joe called me from work on one of those awful wintry days (that's when we were up to our three children) and he said, "What are you doing?" I replied that I was not doing anything and was bored to death. I told him that I wanted to buy a sewing machine and learn how to sew. He thought it was a good idea. He didn't know that I had been researching sewing machines for months. I called wherever they sold the Singers in town and asked the store owner to bring me out the specific model I wanted. So I bought it right on the spot because I knew exactly what I wanted.

Joe was at that point in his life when he became very interested in volunteer work in the sports area. He started the hockey program. He was gone a lot because of his volunteer work. I say to myself now, I was really the old-fashioned mother. He was lucky because I let him go and do all those things. Today I don't know if wives would put up with it. Then what happened as the boys got old enough they got more involved in it and it was wonderful. He really has done a terrific amount over the years for children, including ours. He was a great father and really he still is. Joe was a bundle of energy and he would come home from work and he never said boo about work. Usually I had something to discuss about a problem with one of the kids that he had to take care of, but he would just jump right into something. He changed gears very quickly.

When Joe was home on the weekend it was almost like I wasn't there. I felt like I had the weekend off. I had to still do some of the things, but he was in charge. It was a wonderful feeling. It wasn't even mentioned or discussed; it was just something that was done. It was a natural support system. I don't think the wives would put up with the long hours today. Even marriage was looked at differently in those days; that might have been part of it too. The only way to get a divorce in New York was to prove adultery. We looked longer, harder, and more carefully at choosing a mate because we just assumed it was going to be forever. I know of marriages that didn't last, but I think, really, people stuck together through thick or thin. No matter what happened, you stayed together and supported each other. It was a built-in feeling.

Wives on the Seaway project dealt with many challenges based on their geographic location, numerous children, and unfamiliar surroundings. According to one of the wives, it was her friends who kept her going. These women followed their husbands from one project to another, and because many were far away from relatives, the wives of their husbands' co-workers became their family. The younger women played the role of sisters, while the older ones acted like mothers and aunts. These women, most of whom had never met one another before arriving in Massena, formed strong bonds based on their roles as young wives and mothers. They also gained knowledge and support from many of the older women who had survived many other projects and could impart words of wisdom.

Marge Wiles

I think each of us that came to Massena knew at least one other family and that was such a big help. I made wonderful friends on the Seaway project. I wouldn't trade that experience for anything. We have always been there for each other in the good and the bad. We didn't have much time away from the kids because they were so young, so what socializing we did was with our neighbors in what we called "the development." Sometimes we would have potluck dinners with the kids when they were young. It was hard to find babysitters. We wives traded babysitting a lot during the day. When we did socialize, it was as families. When the kids got old enough, we left them at home and went out and played cards or attended a cocktail party. We never seemed to miss what we didn't have. It was just a very special time. I just can't say enough about the whole bunch of women. They are still very special to me. I just lucked out. There were times when I thought, "Gee, I wished that I didn't live in this fishbowl," but they were very few and far between.

Joyce Eastin

My neighbors and I played bridge and cards while our kids played board games or went outside in the yard. Our husbands all worked on the job, and because we were all one-car families we did a lot together. A lot of the children in Buckeye were the same age as mine and since we lived so close together, I could get together with the other wives and visit, have lunch, or watch their kids. I went to Potsdam one time and one of the ladies from church was surprised that Helen and I drove all that way to go to lunch. It was only a thirty-mile drive. I thought nothing of it. But then, when you live in the West and you have to drive so far to get anyplace, that distance was nothing!

I had a whole bunch of lady friends. Of course we were from all over the country, and we were plopped down in the same area. We all had the same problems, and we helped each other out. We were far away from family, and if my kids got sick and I had to get to the doctor's and I didn't have the car that day, I could call on one of them to watch my other kids or give me a ride. I can't stress enough this one-car-family thing because it meant that there was carpooling, and if your husband drove that day and you didn't have the car and something came up like an emergency with one of the kids, you had to call on your neighbor to help. Everyone was terrific about that. One of my good friends had five kids and she didn't even know how to drive a car. Even if her husband left the car for her, she couldn't do anything with it, so she depended on everyone around her to help her get around. We were all happy to do it because she was such a sweet person.

Phyllis Cotter

My husband had worked at the Hungry Horse Dam in Montana for the Bureau of Reclamation, so we knew a lot of people before we moved to Massena. Many of them including the Buirgys lived near us in the Buckeye Development. Irene Bryant and I used to go to Potsdam and shop and have lunch. We were really good friends. I don't think Irene really liked Massena. She liked her neighbors and liked the people in Buckeye, but I think she felt stifled by her kids and the area. All of us wives would not have been able to easily survive up there if we hadn't had the Development. We all relied on each other for support, babysitting, and entertainment. Unless you could have hooked onto a really good neighbor who was friendly and willing to help you in an emergency, it would have been tough.

One of the things that Helen Buirgy and I still laugh about is how mad she used to get about what a good cook I was. We both had to make lunch for our husbands because there wasn't anyplace to eat. Jim has carried a lunch since he started country school. Every morning I packed a lunch for him and it usually included a piece of chocolate cake. It made Irene Bryant so mad because Jack would come home and tell her all about the great big piece of chocolate cake, and she wasn't about to make him a cake just so that he could put it in his lunch.

Ann Marmo

Joe and I still keep in contact with the couples we met there. It was like a built-in family. It was certainly a wonderful group of people. Joe and I had a big house with big rooms in it, so we had many parties. We got to know these people pretty well because every time they needed a place for a large gathering, they used our house. We were all people who had moved around and gotten away from our families. That is so common now it is hardly mentioned, but in those days it wasn't the norm. But we realized after the fact we became more independent. I think what resulted were these deep friendships, and we took care of each other. We really literally took care of each other and each other's kids. I was involved in several bridge groups. We also bowled with Joe's co-workers and their wives. I used to spend all day at home looking forward to doing something in the evening.

Many of the wives of the Seaway and Power Project workers have fond memories of their years in Massena. Unlike their husbands, who spent the majority of their time on the various construction sites, these women had to deal with local residents, the educational system, and boredom. Their reflections concerning their time spent in Massena mirror the daily struggle of many women in the 1950s who were away from their families and dealing with the challenges of maintaining a stable home life and downplaying the desire to have a career outside the home. From 1954 to 1958 these wives and mothers all shared an experience that challenged the strength of their marriages and their survival and social skills. Much like troops that fought on foreign soil, these women were transported to a strange

30. Joseph and Ann Marmo. Courtesy of Ann and Joseph Marmo.

land, to live in harsh weather conditions, among a hostile population. Fifty years later many still keep in contact with one another and often reminisce about the time they spent in northern New York.

Marge Wiles

One of my fondest memories is that my husband and I won a marvelous, all-expense-paid trip to New York City. There was a little dairy right across the street from our house and the owners were sponsoring a contest. My oldest son came home one day and said he had signed up for it. Luckily, even though he was officially too young to enter the contest, he had the same name as his father. Several weeks later my telephone rang, and when I answered it, the contest organizers were on the line. I had to answer the question they asked me correctly. They had five musical notes on a magnetic board which I was watching on my television. I had to put them in the right order so that when they were turned over it would spell music. The game was called "Match the Music." I had been watching the show for a long time and I said, "If they ever call me, I am just going to say reverse it to five, four, three, two, one"—and by golly I won!

When people found out I had won, the offers to help us with babysitting or any other assistance we needed just started pouring in. Our friends were all so excited for us. Jeannie, our youngest, was just a baby and I farmed her out to one of our neighbors. The other three were cared for by different families. It was in the dead of winter and how we had the nerve to do it, I don't know. We drove to Ogdensburg through deep snow and then we came back in a blizzard. We flew on a puddle jumper of a plane and when we come back to Ogdensburg, our car was buried under four feet of snow, but we had the time of our lives in New York City. Everything was paid for. My biggest regret was that we didn't make it back with the kids.

Another exciting thing was seeing the many dignitaries that came to visit the area during the project. I can remember watching Senator John Dulles get off his plane and President Harry Truman drive through town. There was a guesthouse a block away from my house where people like that stayed. During the dedication Eisenhower and his wife, Mammie, came and I was standing on the side of the road when his limousine

31. This May 1959 photo shows the heart of the St. Lawrence State Park Recreation Area. Barnhart Island is in the foreground. Long Sault Dam is at the upper right and shipping in the Seaway can be seen in the upper left. Courtesy of Alfred Mellett and PASNY.

passed by. He just waved at me like we were old friends. At the ceremony at the Eisenhower Dam, I saw Queen Elizabeth, Prince Philip, and the Rockefellers. It was just a fantastic experience for all of us. We were sorry to leave.

Barbara Hampton
It is wonderful to think back at some of the crazy things that happened. Massena is not a very racially diverse city. This was true of a lot of small northern towns in the 1950s. Before the civil rights movement, most of the blacks lived in the South or the larger cities. The most unusual experience I had when I was in Massena was when my family and I were leaving Million-Dollar Beach. Sometimes my neighbors and I would take our

children to the beach while our husbands were at work during the week. On that particular day Bob was with us. As I was walking along I looked to my side and I thought I saw a shadow and then I looked again and there were some very tall people standing there. They were not like the people today who call themselves black. These people were as black as an absolutely black piece of cloth. It was hard to see them. I believe there was a man, who was quite outstanding looking and I thought rather handsome, a woman, and a teenager. We just stared at them when they walked by because we had never seen anyone before that looked like that. I turned to Bob and said, "Look at those people, isn't that unusual? Where do you suppose they came from?" We found out later that they originally came from an island off the coast of India. I don't know if they were working on the project or just visiting the beach from Canada.

A black lady also came to my door one time. I don't remember if she was selling something or if she was a missionary, but in all the time I lived in Massena, I never saw a black person except at the beach that day. It never occurred to me that neither had my daughter. When I opened the door, Julie, my two-and-a-half-year-old daughter, started screaming. This happened very early on in our stay in Massena and I took pains to educate my daughter about people of different races and cultures when I brought her to see my parents in Albany, New York.

Joyce Eastin

I suppose at the time it may have been difficult, but I met an awful lot of wonderful people every place that we have lived. You were really isolated in the wintertime, but my kids loved it. It was a good steady job and Howard loved it. It is very definitely a way of life. None of us would have survived if we had not had one another. My closest friends were three and four years older than I, and my next-door neighbor was nine years older. We had children really close to the same age. It was a fabulous change. It was a big change for me. I was living around women and families who had the same job that my husband had. I had always lived in town before. The friends I met there I have kept in touch with and we have remained very close. It really was a good experience. I wouldn't trade it for anything. It was a very good time and we met so many wonderful people.

Phyllis Cotter

Some of my friends did not look at their experience in Massena in a positive light. Irene Bryant was stuck at home with small children like I was. She was not an enthusiastic mother. She was more of a modern woman and was mad as a wet hen that she couldn't start a veterinary practice. Ann Marmo and Joyce Eastin weren't too happy there either. Their interest in outside activities was less than mine. They didn't like traveling around to the historic places and didn't appreciate the natural beauty of the river and the nearby mountains. I am sure for the wives to go out and get a job anywhere at that time would have been very difficult because it was not the norm, especially in Massena. I was offered a job as a part-time medical lab technician at the local hospital by my doctor, but I was not interested in working when I had small children. He did give me a professional discount when I delivered my son. Everybody was different, but we are all still good friends and can now laugh about our time on the Seaway. The one thing we agree about is that we are all thankful we met each other.

Helen Buirgy

I don't have many good memories of living in Massena. My little girl, Joanie, got hit by a car and was very nearly run over. She was not badly injured, but sustained a black jaw and a chipped tooth. The young man who was driving was cutting across on our street to get to work. He stopped when he saw my daughter running across the street in front of him, and in fact, if he hadn't it wouldn't have been a good story. He stepped on the brakes right away and the forward motion of him stopping is what bumped her. The reason she was crossing the street was to meet her sister who was fourteen months older, who was getting off the bus after her second day of kindergarten. I was confused over which direction the bus was coming from because the driver switched his route every day. He had come from the north the day before, so Joanie and I were waiting over there. The bus came from the other direction and I didn't take hold of her hand. I looked left and right and then left again and I saw a car coming and she had already gone over to where her sister was.

32. Residents of the Buckeye Development, including Helen Buirgy and Mrs. Kenneth Hallock, taking a boat ride through the completed Seaway and the Eisenhower Lock. Courtesy of Kenneth Hallock.

When my husband got home from work he heard the story of what had happened and the young man was still waiting for the police, but they never came. Our road was not patrolled by the county sheriff or the officers from the Massena Police Department. Ralph finally asked the driver to move his car out of the road because obviously the police were not coming. After the accident, they started patrolling the street because the Buckeye Development became part of the city.

Irene Bryant

I worked very hard to be where I was and at that time that was very unusual. I got my degree in 1947 from Washington State Veterinary School. There were three women in my graduating class. I worked two years in that area before I got married. Women received the same training as men did. I would have been happier working outside of the home even after I had children. However, I had a mother-in-law who didn't believe in women working outside of the house until their kids were grown. I don't think she ever wanted me to be a professional.

The only contact I had with the local people was with the local veterinarian and his family. They were very nice. I had my New York State license and the local vet did not know how to do ear trims and back then the owners of boxers, Dobermans, and Great Danes had their ears trimmed. So I would do those for him, but that was about it.

I was one of these awful wives in the 1950s who followed her husband around. I didn't question it. I don't think wives made demands on their husbands or argued with their decisions; it just wasn't done. We were the docile wives. Were the other wives happy? You see we didn't discuss it with each other. The friends kept you going. I had a close group of friends within the development. We entertained ourselves and a lot of us knew each other from previous jobs. It was not a good place to be. You just existed, you really just existed. I had my friends and attended parties, but I didn't really develop. I stopped living. Living in Massena, I had a feeling that on the big river life was flowing, but I was in the backwaters there. Those were not the finest times for me. We saved our money there so Jack could go to Cornell Law School and that worked out well.

Ann Marmo

Living almost anywhere at that time would have been tough for me because raising children then was so different than it is now. Women were so much more confined. Most of us wives were very interested in moving forward and progressing. Many of us wished we could get out of the house and pursue careers.

I had two very close women friends and we spent a lot of time together. I realized sometime after the fact that we were feeding off each other. We were complaining to each other and comparing notes. I think that if we had not done that we would have been better off. We would come up with reasons that the other ones had not thought of about why we didn't like living in Massena. One of the local gals I met after I was up there for a while heard me complaining about something and it wasn't the first time she had heard my negative comments. She said, "Ann, if you like Colorado so much why don't you go back!" Well I would have in a heartbeat if I could have. So that was kind of a wake-up call for me. I thought, "I can't keep doing this. It doesn't do any good." I'll tell you, though, when we got in the car to leave I said, "Thank God we are finally leaving here!" That's what my thoughts were. I was so relieved. It was fantastic experience for the guys, but evidently not for their wives.

8

A Lifetime of Memories

THE AMERICAN PUBLIC WORKS ASSOCIATION heralded the St. Lawrence Seaway and Power Project as one of the top ten construction projects of the twentieth century. However, even with this recognition, it remains absent from most history textbooks. Students are taught the impact of the Panama Canal and the Hoover Dam on international trade and power production, but cannot locate the Eisenhower Lock on a map of North America. Even though the St. Lawrence Seaway was the largest inland waterway ever completed, and the Robert Moses–Robert H. Saunders Power Dam sends electricity to many areas of New York State, Vermont, and the provinces of Quebec and Ontario, ships sail through the locks and the turbines rotate in relative obscurity.

One of the main reasons why most Americans and Canadians do not know about the St. Lawrence Seaway and Power Project is because while the construction techniques and design may have been innovative for the 1950s, today they are obsolete. In the present, the system struggles to serve its navigational and power production role in an age when people barely concern themselves with current events. The Niagara Falls project, which was completed in 1964, upstaged the St. Lawrence project in terms of size and public visibility.

David Flewelling

Most Americans do not know what the St. Lawrence Seaway and power dam is. I know about it because I worked on it. For the average person that lives in the Northeast they have no idea where their power comes from. Young people know there is a wire connected to a pole in their front yard that transports electricity, but they do not know anything about the dams that generate the power or how these facilities were built. It was memorable for me because it began my career working on heavy construction.

Roderick Nicklaw

The Seaway and power dam was one heck of a project. It was a big deal. It was like the Panama Canal or the Hoover Dam. For the average citizen to understand the magnitude of it, they have to see how the whole system—the locks, dams, and waterway—function together.

Ted Catanzarite

The Seaway and dams haven't been of major interest to Americans in the past or in the present. While it was under construction many tourists came to visit the area. But even though it was one of the greatest construction projects in the history of the United States, in the present day it is largely ignored. Americans should know about the long political debate surrounding the funding of the construction and the modifications that were made to the lock and dam designs over time. All of the elements—the machines, the workers, and the stringent time schedule—were unique. The other issue may be the isolated location of the dam and locks.

Bill Spriggs

The reason why no one knows about the Seaway construction has to do with the lack of American interest in the nation's basic history. People in general are not curious about any event or past project that doesn't affect them. PASNY and the Seaway Authority had the opportunity to build museums on the lock walls and show how each structure was built. But they were cheap. They just wanted to build the visitor's center at the main power dam. If the tourist facilities had been more elaborate, it would have attracted more visitors to see not only the large machines and the unique construction techniques that were used, but also cannonballs and oddly shaped boulders that were discovered when workers dried up the rapids with the cofferdam and dragline operators dredged the river bed. The National Park Service constructed facilities like that at the Hoover Dam. It

33. In April 1958, crane operators removed the cells of Cofferdam C-1, which had prevented water from flooding the main power dam site since 1955. Courtesy of Alfred Mellett and PASNY.

was amazing what all of the contractors and workers did, but their efforts have largely been forgotten.

Joe Marmo

In those days, I don't think people thought much of the historical significance of the project. Everyone finished their part of the job and took off. But it was such a novel undertaking with the massive equipment like the "gentleman," the harsh temperatures, the drying up of the rapids with the tetrahedrons, the land acquisitions, and the round-the-clock work schedule. Those were all unique aspects of the St. Lawrence project and are not typical conditions or problems contractors or workers would encounter on any other mega project. But for those of us who worked on it, addressing contractors' complaints and finding solutions to unexpected problems was part of our everyday lives.

Roy Simonds

I don't have a real inclination about why it is not as well known as the Hoover Dam and the Panama Canal. I often share with people my memories of working on the Seaway construction and most of the time I realize that they have no idea what I am talking about. I don't know what caused the next generation to not be informed of the importance of this monumental undertaking. I really thought that the power dam and the locks were going to be big tourist attractions.

About three years ago I went back to Massena and it hadn't changed a lot. I drove around and saw all the dams and the locks. I would have expected the power plant to be teeming with tourists and there was hardly anybody there. I thought the most impressive thing was when I happened to drive through the tunnel underneath the Eisenhower Lock at the exact time that a ship was being lifted. I was proud to see all of the facilities that I helped build still in full operation.

Thomas Rink

It is amazing that no one knows anything about the St. Lawrence Seaway and Power Project. The project was very complex. It took a lot of legal work, a lot of planning, and a lot of manpower. At the time it was the biggest construction project in the world. I talk to people now and tell them I worked on building the Seaway and they say, "What was that?"

Glenn Dafoe

When I think of the magnitude of the Seaway and Power Project and considering it was built almost fifty years ago, it was really something! There had been nothing of that magnitude before. The only thing I could compare it to was the original canal system built on the Canadian side from Montreal to Fort Culvert. The ability of the project planners to harness the power of the Long Sault Rapids with the removal of all of those large boulders and the swirling current and tremendous undertow still amazes me. I remember sitting on Sheik's Island as a kid and being thrilled to be at the head of the rapids. All of that natural beauty is gone, but what many people don't realize is that it provides power to many areas of Quebec and Ontario. I was proud to be a part of a project that has had such an important impact on people's lives.

Garry Moore

In terms of the general public, the St. Lawrence Seaway and Power Project has been forgotten except in the new towns of Long Sault and Ingleside, where there is a museum whose curators try to keep the history of the Lost Villages alive. But basically it made no impact on the next two generations. Since it was built, Canadians have had no reason to think about the power dam or drive a boat through the Seaway. As far as they are concerned, these facilities have not made a difference in their lives at all, so nobody knows anything about them. However, the American Public Works Association voted it one of the top ten public works projects of the twentieth century.

Frank Wicks

Once the locks and the dam were done, people forgot all about them. They were just there. However, from 1954 to 1959, the Seaway and the power dam were the world's biggest construction projects. Residents of Massena and the surrounding towns were really excited that the construction workers and their families were going to bring financial prosperity to the area. Growing up in Canton, New York, I didn't realize I was living in a depressed area. I remember even before the Seaway legislation was passed, my grandmother always talked about all the great things that were going to happen. Younger people also shared this enthusiasm. When I was on the high school golf team, we played a match at the Massena course. Members of the opposing team joked about the fact that to make way for the Seaway, contractors were going to have to flood the existing golf course and build a new one. Eventually they stated they would go fishing over the old golf course with their kids. In 1955, the typical Sunday afternoon drive was to go to Massena and get an update on the project.

Donald Rankin

At the conclusion of the project, my family and I went to Florida for a vacation. Having been absorbed in the Seaway and Power Project, it had been a preponderant part of our existence and we expected the rest of the world was waiting out there to hail and clap for us. I was fairly surprised that the rest of the world was going on in more or less its own way, and I decided that in the scheme of things, it was not all that important. It was a very big project. But it was an era of big projects. At the same time that the Seaway and power dam were being built, the owners of Alcan were erecting a big aluminum production plant in British Columbia. There were also other big public works projects like the Mica Dam.

Les Cruikshank

The reason why many Americans do not know anything about this project is because the locks are not the biggest or the highest in the world. Even today there are no records set in terms of tonnage during the short shipping season. Also, northern New York and eastern Ontario both have something in common: the areas are not very heavily populated and are at the back end of just about everything. When I go across to see my buddies in Waddington, New York, they say that the leaders in Albany like to keep this area green and do not provide any incentives for company owners to locate facilities there. They are paving over everything in the southern part of the state, so they want to keep it green up here. Politicians seem to want the entire region to remain rural and economically depressed.

Even though the St. Lawrence Seaway and the Robert Moses–Robert H. Saunders Power Dam are not heralded by contemporary members of the general public as construction feats or modern wonders, from 1954 to 1959 tourists and world leaders flocked to the area to view the large machines and the massive earthmoving and concrete-pouring efforts. Some world leaders had a keen interest in the design of the waterway and hydrodam because they planned similar projects in their nations. They anxiously spoke with contractors and engineers about weather-and design-related dilemmas. American presidents and Canadian politicians visited the area to view how their funding was being spent. Members of the International Joint Commission (IJC) also frequently sent engineers and political appointees to ensure that the agencies involved were not violating their waterway regulations. Contractor representatives escorted official guests on site visits and agency representatives often organized luncheons with local dignitaries. Many of these visitors would not have otherwise traveled to Massena or Cornwall.

Arthur Murphy

As part of the media relations department of Ontario Hydro, I was in charge of organizing the transportation for visiting dignitaries as well as any dinners and social events that surrounded their stay. In this capacity I met many politicians, including the chairman of Ontario Hydro, Robert H. Saunders; Canadian Premier Leslie Frost; the Canadian ambassador to Washington, Arnold Hainey; and Queen Elizabeth II. An incident that happened with Arnold Hainey is probably my most memorable experience on the power project because I almost got fired. My boss wanted me to organize a tour of the various construction sites for Hainey and his entourage starting around Prescott in a tugboat. Hainey needed to inspect the shore on both sides of the river for signs of erosion, so he could prepare a report for the IJC, whose members were responsible for monitoring the water flow and controls in that area. It was my role to coordinate this kind of thing and arrange for lunch. So I met with the owner of the King George Hotel and Restaurant in Cornwall, to decide on the menu. I told him that I needed one of his banquet rooms for two hours where my thirty-five guests could enjoy a nice meal and some wine. He took me down into the hotel's basement where the wine cellar was located and showed me a bottle of wine that he highly recommended, so I signed off for him to serve several bottles.

The day of the event, the site tour went off without a hitch. Then Arnold Hainey, myself, the mayor of Cornwall, and other local dignitaries sat down for a nice lunch. Hainey got up and commented on how wonderful the day had been and how he had enjoyed the company of his American counterparts. Then he said, "I have been treated in Ottawa and I have been treated in Washington, but I have never had such good tasting wine in all my life!" Little did I know the wine cost about five times the price of the meal. I started to panic when I saw the bill for the food and wine because I didn't know how I was going to pay for it.

Alfred Mellett

When the construction was going on, I met many VIPs from all over the globe who had come to view the biggest construction project in the world and I took their photographs. I talked with presidents, royalty, and dignitaries from other countries, so it was really an interesting time for me. I also was the official photographer at the dedication ceremony, so I have the best picture of Queen Elizabeth II and Prince Philip with the Nixons and the Harrimans in the background. It represented the cooperation between the United States and Canada and the international significance of the waterway and power dam.

Joe Marmo

Another thing I remember was some of the super visitors: Queen Elizabeth II, Vice President Richard Nixon, and President Harry Truman. At one point the Power Authority recruited me to be a tour guide. People flew into Montreal because that was the biggest airfield. Dominic Nicola drove up there to pick guests up and brought them to the project. Also, because we were so close to Washington, D.C., and New York City, a lot of common people visited the area.

John Moss

It was unheard of back then to have a construction job of that magnitude, so that created a lot of interest. The construction sites became big tourist attractions. Ontario Hydro funded bus tours led by their own trained guides who knew all the political history of the project as well as the design and construction aspects. The guides would stop at the various sites and explain what was going on and what dam or lock would be there when the project was completed.

Roy Simonds

I met a lot of high-ranking politicians who came through to visit the job during construction. It was the biggest project in the world at the time and people from around the world were interested in what we were doing. I often saw the tour guides showing tourists around. Many of the important dignitaries stayed at a guesthouse owned by PASNY which was near my house at the Buckeye Development.

Bill Spriggs

I got a once-in-a-lifetime opportunity to dine with Prime Minister Diefenbaker, the president of the United States, and the queen of England in Montreal. I was in town during the opening of the Seaway and I was invited to attend the dedication luncheon. I would not have met these people otherwise. I don't think many of them knew where Massena was before that day. There is a famous story that when Lionel Chevrier went to visit President Truman in Washington about the need for the joint construction of the Seaway and power dam, Truman leaned over and told him that he really had to tell him the location of this town called Massena.

Most of the workers who completed the St. Lawrence Seaway and Power Project never forgot the time they spent on the job. In general they cherished the opportunity to work on the Seaway for similar reasons. Firstly, the contractors paid them higher wages than other employers had offered them in the past. Secondly, many funded their college educations with their earnings from the project, while others started saving for a first home or eventual retirement. Many carpenters, electricians, and laborers received on-the-job training, which led to future job offers on other waterway and public works projects. Finally, some workers pursued careers in civil engineering and law because of the engineers and attorneys they interacted with on the Seaway project. Therefore, working on the Seaway and Power Project had a lasting impact on their lives.

David Flewelling

The Seaway with all of its locks and dams opened up my eyes to what civil engineering was all about. Not only was I paid well, it was also personally rewarding work. I knew I was doing something that would last a long time, and as it turned out, I worked on multimillion-dollar projects most of my life. I don't think there really has been anything that has approached the magnitude of the Seaway. There was work all over Massena, from the

Intake area down through Long Sault, along the Grasse River, and for fifteen miles downstream. It was a big project. The dredging and lock and dam sites were so spread out. There were sites near Waddington all the way to Montreal. The amount of detail and planning that went into each particular site was awe-inspiring. It just opened up my eyes to what was out there and what could be done in terms of construction and taming natural elements.

Ambrose Andre

It was my first job fresh out of college. I was seeing equipment that I had never seen before. We didn't have television. We didn't watch the Discovery Channel. They brought walking draglines up from the Kentucky coal fields. There were huge off-the-road trucks with tires that were six feet in diameter. The equipment was very impressive. It was all the kind of construction stuff that I enjoyed, blasting and machinery. At that time it was about as big as it got in New York State or the world. I still just marvel over the complexity of the project. The fact that the power dam, the locks, the channel, and the bridges operate without raising the water level is fascinating.

Frank Wicks

The Seaway project played an important role in my life at that time. Being a member of a concrete crew gave me the first sense of what was involved in completing a construction project. This experience piqued my interested and I later pursued a doctorate in engineering. The months I labored on the Seaway project were an important step in my education and development. It was forty-seven years ago, but my memories of it are still quite vivid. I definitely had a sense of pride working on it. The machines that I worked around and the height of the locks were impressive. The Euclid trucks and earthmovers were enormous.

Garry Moore

Every day I worked on the power dam, I felt like I was doing something that hadn't been done before. I thought it was very creative, unique, and very innovative. It was almost like the space program. It set the course for a lot of people's lives. It was good pay and allowed many of us to buy homes and start saving for our retirement. I didn't need a lot of money since I was living at home, but I survived on my overtime and banked my regular check. I was supposed to work eight hours, but I worked eleven so that was three hours of overtime a day for which I was paid time and a half. It was also the beginning of my association with the Seaway and power dam. After the Seaway was completed, I was hired by the Seaway Authority as a civil engineering technician in Cornwall in the operations and maintenance department. I took care of the Iroquois Lock, what was left of the Cornwall Canal, and the two locks at Beauharnois. I had been writing exams for several years and I finally got my diploma in engineering in 1967. I was then offered a position at the head office in Ottawa, which I held until my retirement. It is a shock to the system to remember all of these old memories. It was a positive experience.

George Haineault

Working on the Robert Saunders Power Dam set me up in a trade for life. I was exposed to new and large technology, so it was quite an awakening. It was sort of a boot camp for me when I was there. It was dangerous, but it was very exciting, a real learning process. I spent most of my time working in one hell of a big hole. When I began working there, my supervisor would order the operators to move over a little bit of dust and he would say, "We are going to build a generating station on this baby!" They continued to dig until they hit bedrock. It was unbelievable how far they went down. I had to get the hell out of the hole at the start and end of my shift because once they hit rock, they were blasting. Before they set off an explosion, the foreman would blow one quick warning whistle. I also remember that I never had trouble finding cold water to drink because it was surfacing out of the main. The water had some sulfur in it, but it was cold, a good drink. The experience was so exciting and so different, I could almost relate it to the way folks who settled the Yukon must have felt.

Arthur Murphy

I think being part of the St. Lawrence Seaway and Power Project was more memorable to me than other Ontario Hydro projects I was involved with during my career because I was a very young man and very impression-able at the time. It was my first job with the agency and the nature of my work was very personal and very topical. I also could relate to the anguish of many of the landowners whose property was flooded because my grandfather's dairy farm in Dickinson Landing where I had spent a good part of my childhood vanished. I also met my lovely wife. So, a lot happened in my life during the time I was involved with that project. After the St. Lawrence Power Project, Ontario Hydro constructed hydrau-lic plants up north in the James Bay area and then primarily focused on building thermal and nuclear plants.

Alfred Mellett

National Geographic did a story on the St. Lawrence River and the construc-tion using my photographs, and later they hired me. They did two large spreads combining my photographs with those of their staff photogra-phers. After that project was done in 1960, I went to Washington. My expe-rience with the Power Authority set me up in a career that allowed me to travel to India and the Middle East to photograph and work on other navigational and power projects. I never would have had that opportunity had I not come to Massena.

Bill Spriggs

It was a glorious job. I really enjoyed every minute of it. I was there without portfolio and given a lot of responsibility handling the electri-cal inspections on all of the Corps' sites at the same time. In the Corps everybody worked with everybody. It was like we were birthing a child and then watching it grow and flower. I was there when the Eisenhower and Snell Lock sites were just flat pieces of ground and some of the trees and the earth were being moved, and it grew into two cement boxes with water in them. Just to see all of the excavation, concrete pouring, and door installation completed and operational from my perspective was something I have never forgotten. Given the skills I had at the time, being able to be involved in something like that was a gift. It was amaz-ing what we did.

Roderick Nicklaw

I remember the work ethic in the 1950s was different. Back then it was a day's pay for a day's work. People didn't think the world owed them anything. You had to work for it and if you didn't you got your butt fired. Men didn't start at the top. They began their careers at the bottom of the ladder and worked their way up. I worked with a bunch of nice guys and I made good money.

The only thing that really bothered me on the Seaway project was the waste, in terms of throwing away welding rods, plywood, and all the other building materials. The welders would bring up a box of welding rods when they were putting the rebar together for a pour and when their shift was over they would just leave the whole box there. If it got wet in the rain, they would just throw it over the side and backfill over it. The carpenters also burned enough plywood to construct half of the buildings in Massena because if the concrete got on one side of it when they did a pour, they threw it out. It was ridiculous. My boss said it cost him more to clean it up than to use new plywood. Also, if a form didn't work because it was built to the wrong specifications, they would just bury it. The operators used to push the wood into great big piles with their bulldozers and then burn it. The workers disposed of millions of dollars of perfectly good building materials. This practice takes place on most construction jobs, just not at that level.

John Moss

First of all, the reason why I remember working on the Seaway was the size of the thing. Everything was so big, the machines and the enormous concrete structures. I also made good pay. The Ontario Hydro employment office and their subcontractors offered all of these jobs to local people and that was quite a major thing to happen in a small area like Cornwall. They also brought in many skilled workers and their families from all over Canada and the world. The local industries were not very happy with Hydro paying big money, so they had deals with Hydro that they wouldn't pay over a certain rate. I think the reason many of the single men stayed for the duration of the project was the good-looking girls. Many of the local young ladies were very friendly and looking for husbands. The fact that the men were working and the money was fairly good was enough to keep the ball rolling.

Thomas Sherry

I still remember the power dam as the biggest job I ever worked on. I had been employed on a lot of other jobs as a mason. It was like working on the Panama Canal. There were giant trucks with dump boxes that held three regular-sized dump trucks. There was a gantry crane on railroad tracks that was used to place concrete in the forms. The aggregate in the concrete was like boulders, instead of the usual number-two stone used on most jobs. They were ten times larger than in ordinary concrete. The crane operators would pick up a truckload of concrete in a big clamshell and then open it up and it would pour right out. Everything was on a big scale. Everything was like taking a small job and magnifying it ten times. I never saw anything as big as those trucks and earthmovers. They were monsters. The tires were higher than my head when I was standing next to them.

Ted Catanzarite

When the Seaway construction finally began in 1955, I was lucky enough to get hired by one of the contractors. It happened just perfectly for me. My father had died when I was ten and this was a way for me to save money for college. I am not sure what I would have done otherwise. I earned enough money to cover my tuition of $500 a semester. I still would have gone to college one way or another, but it made it easier.

Thomas Rink

It was a good education for a young guy. I didn't make or save a lot of money, but I learned what the world was about. It was awesome for me. I was a young kid and I was a hard worker. I remember I went back to Hornell after I had been in Massena for about two and a half years and my dad asked me how much money I had in the bank. I told him that I had bought a car on credit for $600 and was spending the rest of my paycheck on rent, food, and beer. He asked if my goal in life was to fill the pockets of the owners of the beer joints, or if I wanted to get married and buy a house? I will never forget that. That conversation with my dad woke me up a little bit. So every week after that, I saved a little money and when I got married I had enough for the down payment on a house. The money I earned on the project gave me a financial head start.

It was the beginning of life for me. I was very active. I can remember the old guys telling me that they went to bed at 9:30 because they were tired. I would get in bed about midnight. I would get done bowling around 9:30 and go and have a couple of drinks. I didn't care if I only got five or six hours' sleep. That was all I needed. It didn't bother me. But if I overindulged it might bother me for a few hours in the morning, but I shook it off. If they had a breathalyzer test at the gate it would have been tough because there would have been a lot of people sent home. I never saw any drinking on the job. However, when my shift was over, I had a few beers and if I got hungry I went down to the restaurant for something to eat and then I went home and went to bed. I got to know the people I worked with and then they moved on to another job and sometimes I never saw them again.

Kenneth Hallock

The Seaway and power dam was the biggest project I worked on in my career. I probably remember it more than others for quite a few reasons. Firstly, I was very young in the 1950s and not very high up in the Corps, but it struck me that I had the opportunity to work with many of the top engineers and administrators. I also got entangled with the Seaway Authority because of the concrete problem several years after the construction was completed. It was the only major project that I was involved with for seventeen years. Usually I helped design a waterway or lock project and then it was put out for contractor bidding. But in the case of the Seaway, I was part of not only the design of the locks, but their rehabilitation. I saw my plans completed and I lived the whole thing through. In the 1970s, I applied for a job with the Seaway Development Corporation as superintendent of operations, but I didn't get it. John Adams, another former Seaway worker, and I both sent in our resumés and he was hired. It is probably a good thing I didn't get that job. I have said so one hundred times. This isn't meant to be derogatory, but the Seaway Corporation is a much more political animal than the Corps of Engineers, and I would not have been comfortable in that environment.

The men who worked on the St. Lawrence Seaway and Power Project spent four years on a project with global impact and a diverse workforce that changed their lives in a positive way. Engineers and laborers interacted with contractors and photographers and workers from various provinces and states they otherwise would never have encountered. One would have anticipated bar fights, strikes, and a strained working atmosphere based on different cultural values and inherent prejudices. Instead, this continental crew managed to learn from each other and appreciate their differences. Many Massena-based workers had never met a southerner before and were fascinated by their stories about their past work experience and contrasting lifestyle. Several of the skilled workers and engineers had served in the armed forces during World War II and the Korean War, and compared the camaraderie with their fellow workers, the struggle against time and weather, and the urgency to complete the project to fuel new manufacturing plants to the bond they had formed with their platoon members during tours of duty in the foxholes in France and on the Yalu River. To them, working on the Seaway and Power Project was a once-in-a-lifetime opportunity to serve their country and to build a navigational and power facility for the betterment of future generations. The most experienced planners and contractors on both sides of the border set aside their egos and past differences to complete the most massive and complicated power project ever undertaken.

Robert Carpenter

I worked with many men from Tennessee and Alabama. They had different cultures and beliefs than me, and they talked different. It took me months to completely understand what they were saying, but they were great guys. They had worked on TVA navigational and dam projects and had a lot of tales to tell. I enjoyed listening to them try to outdo each other in their thick southern accents. I worked with the cream of the crop from all over the country. Many of the local people could not operate equipment or conduct inspections, but gained employment as laborers or electricians.

David Manley

Working as a laborer on the Seaway construction sites was actually the first real job I ever had. I used to work on farms and did odd jobs, like mowing lawns and shoveling driveways. I was just awed by it. Some of the equipment was so big. I hadn't seen big trucks and gantry cranes before. I also met all different kinds of people. That really stuck in my mind. For the most part people got along. Every once in a while men got in an argument or a fight. Before that everybody I knew was from the Malone area or Massena. Some of them told stories about living down South and of course this was all new to me. People had different ways of living and, being from a warmer climate, they talked a lot about the weather. Back then they also talked about the colored people. We did not have too many black people on the project. The southerners talked about how they treated them. People from Georgia and Mississippi said the black people knew their place. Of course they didn't use that language back then. I thought they were pretty nice people. A lot of them had moved all around.

Joe Marmo

One of the things I remember was how all of us got along. It didn't matter if someone was from Alaska or Massena or had a doctorate degree. I worked with a lot of skilled engineers, carpenters, and machine operators. People came from all over the world, including Bob Coke, who had worked on hydro projects in South America. Jim Hendricks from Tennessee had the biggest southern accent I had ever heard and his lifestyle was different from mine. But we had to work together. Nowadays if you go into an established office you have to deal with office politics and find out how to fit in. But up there, we all managed to work together even though we were from different geographical areas. It was such a unique project. Contractors needed men from all of the skilled trades as well as unskilled laborers on the Seaway that they might not have employed on other sites. Workers coming from all over the world, each with their own egos and their own talent, and we had to make it work.

Secondly I had to know who I could trust. I had to be mindful of this while getting my job done, which was unique. I also had to deal with my Canadian counterparts. My friends and I who worked on the Seaway and

Power Project were part of the greatest generation. We had been through hell in our lives already as Depression babies and World War II veterans. Many of us were much more mature than some of our high school classmates and just appreciated a nice job and a good place to live. I think we built a cadre of people who had a lot of respect for each other because we had a tough job. We all had to cooperate with each to complete the Seaway and power dam project on time.

One of the most memorable characters I was associated with in terms of contractor claims was a really shrewd guy named Jacob Goldstein, who was a personal friend of Robert Moses and the legal counsel for the Power Authority. Moses trusted him implicitly. Dealing with contractor issues was new territory for Goldstein, but he was a very smart guy. He was in partnership with his son-in-law and Guy Zemora. The three of them used to come to Massena once a month and go over the changes the engineers and the contractors had suggested. I had about eight engineers working with me and we went over every claim with them and made suggestions on how we thought they should be dealt with. Then Goldstein and his partners would take this information back to their offices and negotiate cost adjustments with Perini and the other contractors.

Once in a while I attended meetings in New York City, which was one of the perks of my job. One time I took the train down to the city with Jacob Goldstein's son-in-law and he asked if I wanted to see a Broadway play. I said, "Do ducks quack?" So he got me tenth-row-center tickets to see *Camelot* with Bob Goulet, Julie Andrews, and her boyfriend, Richard Burton. Even though I was dressed like a lumberjack from Massena, New York, I had one of the best seats in the house and it was one of the highlights of my Seaway experience. My wife was so jealous, she almost killed me. She said that I was not allowed to go to New York City without her anymore.

Jack Bryant

Construction seemed to be a very friendly community. People made friends on one job and saw them again on another project. That was certainly the case on the Seaway project. It was a long four years and of course I was deeply involved in it. I really enjoyed the people and the work. In fact I met some of the PASNY lawyers. I was an engineer and I found in negotiations that sometimes the lawyers representing the contractors would be crackerjacks and they didn't know what they were talking about. I could negotiate circles around them. They thought they knew a lot. But the Power Authority hired an outside law firm that was surprisingly very small. One of the lawyers, Jake, was very Jewish and he really made an impression on me. Before the meetings with the contractor, I would prep him and answer any of his questions about how the design changes affected the contractor's cost. In the negotiations that followed he usually got them to agree to his terms. He was very talented, and I got along with him well and liked him. He was one of the reasons I went to law school after I left Massena because he had one foot in engineering and one foot in law.

Donald Rankin

I think the camaraderie with so many different types of people involved in a large project was stimulating. The St. Lawrence Seaway and power dam was the project I enjoyed working on the most because of its size and the variety of workers and construction techniques involved. After I left Cornwall I oversaw four more projects that were not as big. The construction work that Hydro was carrying on was in most cases fairly fast paced and I would spend about three or four years on one site and then move to another. The pace of the work I think kept people very interested and energized. It was very demanding and people thrived on that. There was always a whole bunch of new drawings to look at. The power project was a very complicated effort because there were so many sites and each one was at a different stage of construction.

Joseph Couture

It is hard to find a bad thing to say about working on the power dam. I loved every minute of it. I liked all the men I worked with and I really liked the freedom to make my own decisions on that job. There were many top guys at the power dam site, but they didn't tell me what to do. When they came to the site for a visit, I would often ask them how a certain thing should be done. They were satisfied with the way I worked and when I asked them to help me out, they would say that I would find my way out of a situation when it arose. It didn't matter if it was a top official from Iroquois Constructors or an engineer from Ontario Hydro, he respected me. My workers also respected me because I was not afraid to fire a worker who didn't want to work hard. The other men were thankful to be able to do their jobs without being held back by someone who was lazy or incompetent.

Glenn Dafoe

The work on the Seaway and Power Project was more diversified than other construction sites. For example, there was excavation, concrete mixing and pouring, form construction, and surveying all going on simultaneously. The design and dredging was meticulous, but other aspects like the concrete work was very repetitive. There was a greater variety of not only tasks, but also skill levels and specialties among the workers. I worked alongside engineers with doctorate degrees and site managers who had earned their stripes on many of the large Ontario Hydro jobs around Canada and we all got along.

I always liked to work with my hands and my trade as a carpenter. I got to hone my skills on a project that offered me a once-in-a-lifetime opportunity to be part of something that made an impact on navigation both domestically and globally. I did not know it at the time, but it was highly improbable that this project would ever be repeated in my lifetime.

Thomas Sherry

It really was one of the most exciting things I did because I felt like I was participating in something important. It was like working on a part of history. A lot of us, like the World War II veterans, are on our way out. I was a lot happier being on the Seaway than being in Korea. I was there when the war started. I was in the infantry and on the ground. Korea will stay in my mind forever. I was there for thirteen months. I went all the way up to the Chinese border. Then we had to retreat. I have a lot of bad memories from that experience. I have better memories of the Seaway. The war memories are not good. They haunt me. However, working on the Seaway was exciting.

Frank Reynolds

I enjoyed working on the dredging every day and I had some good guys working for me that I got along with. The big thing for my wife and me was to come up here and spend three years on the Seaway and then go back to our jobs. We got to fish and visit our families and that was important to us. We also got to witness construction on a massive scale. It was a once-in-a-lifetime experience. On the weekends we used to go over the border to Iroquois and watch them move the houses. What an apparatus they had for that! Even now, every once in a while, I find myself dreaming about the job.

William Rutley

The reason why I remember my time on the project was because I did something different every day. It wasn't a routine thing. It was a challenge every day I went to work. It was also shocking to see the communities around Cornwall change. Most of the homes in Ingleside and Long Sault, the two towns that were relocated, didn't have running water. But after the project was completed, residents had showers and indoor plumbing and other modern conveniences like central heating. The Ontario Hydro property department made the new houses and towns look pretty good. Residents got basements, hot water, and furnaces. After all of the houses were moved, all the new occupants complained about was that they no longer had river frontage. Before the Seaway was completed, people who lived along the old Number Two Highway had water in their backyards and could go swimming anytime they wanted to. Now they have to pay the dam parks to go in and swim and a lot of the shoreline along the river has not been developed. People who had water frontage before should have had water frontage after. On the unused land, Ontario Hydro was supposed to create new parks and beaches, but now it is just untamed fields. It is a mess.

Neil McKenna

There were a lot of comical things that happened. I remember sitting at the Eisenhower Lock near the tunnel one time, and one of the guys from our survey party, Andy Van Slyke, was looking at the opening of the tunnel.

I said, "Andy, do you know what the little box on the side of the tunnel is for?" He said, "No." I replied that it was made so the squirrels and the other animals could walk from one side of the lock to the other without being in the roadway. Andy looked at me and everyone in the truck and said with a straight face, "Boy, the government thinks of everything." Of course the cavity was for conduits.

Jim Cotter

There are several events during the construction I remember. Before the cofferdam had quelled the Long Sault Rapids, a professor from a regional college was going to claim the title of the last person to navigate the perils of the rapids. He chose a canoe as his mode of travel. Local pundits reported he had one foot on the riverbank and the other in the canoe as he descended the rapids. Not to be outdone, two Uhl, Hall, and Rich employees, Joe Vanbleet and Dick Payne, took a motorboat and roared up and down through the rapids without fear. Such is the unmaking and making of history.

Another interesting character was Al Mellett. I don't know if he was an endorsement for bachelorhood. I remember one of his objectives in life was to be a *National Geographic* photographer and he achieved that. He told me that one weekend he went to New York City and saw a sports car in the showroom window and he went in and bought it. That's the freedom he had. Another one of his accomplishments was that he was Robert Moses's closest friend in Massena. He dined with him whenever he was in Massena and accompanied him on many trips to New York City.

I departed with my family from Massena for Niagara Falls in 1958 to work on the project to capture the power potential of the Niagara River. Life changed for us and our many friends we had followed from one major hydro project to another when the sites were developed. However, we left Massena with each of those years packed with memories. Our family trips to Upper Jay, Fort Ticonderoga, Vermont, and the Adirondacks in the fall when it was awash in colors are some of the delights of those years. Our youngest son was born in Massena and he proudly proclaims himself a native New Yorker as did his great-grandfather. From our family research, we know there are still Cotters in the area of Franklin and St. Lawrence counties. The earliest ones are interred in the tiny graveyard surrounding St. Mary's Church in Fort Covington. To stop and recall and reflect on the good times and work in the past are treasures available to everyone. There were accomplishments for which we can all be proud and the nation benefited.

A minority of workers would rather forget the time they spent in Massena. These men tended not to be engineers and were few and far between. Financial loss and loneliness based on their separation from family members fueled their negative attitudes. Workers who arrived in Massena without their families often remained in the area for a few months at a time for the entire four years of the project. They found employment with one of the numerous contractors and earned high wages. Others opened businesses and hoped to share in the wealth that many political leaders had publicized in the media. Instead, some of them lost money or returned home with damaged equipment.

Ted Catanzarite

When I worked at the power dam I knew a laborer from Watertown who was also of Italian descent. Because of the tragic nature of his circumstances, I prefer he remain unnamed. He was convinced that Massena was going to expand and he wanted to go into business. This man had no interest in going to university, so he was destined to be an unskilled worker all his life, unless he was self-employed. So he worked on the power dam for a while and then he quit and opened a grocery store which failed a year after the Seaway construction ended. Unfortunately, he committed suicide and that was one of the few unhappy occurrences. His wife was also from the North Country and her father was one of the first psychiatrists in the area. Many other men purchased restaurants or bars which thrived for a few years and then closed when the Seaway workers left and the local population and demand returned to their previous levels.

James Romano

I don't have many good memories of working on the Seaway project. Every time I went there, I came back broke. The reason why I left Waddington was that the Iroquois Dam was the only site still under construction. The dam floor was flooded and the truckers were driving in almost a foot and a half of water. When you drove through that stuff, the water and silt got into your brakes and after a few months you had to buy all new brakes and bearings, so the money that you made went into repairs. It didn't pay, so I said, "To hell with it. I ain't going in there." I also remember I had a fear that when I was down in one of the dam sites the wall might get washed away. When they built the intake and the dams, even though the side walls were made of fifty feet of concrete there was water on the other side. I used to think, man, if that thing ever lets go, I would be swept away just like the characters in a comic book. So when I saw that the work was dangerous, I wouldn't do it.

I left the Seaway in 1959 and as I crossed the city limits, I spat on the sign that said "Massena" and said, "I ain't coming back no more!" I made it back to Rochester in a few hours. When I got up the next morning, I looked at my truck and there were broken springs in the back from the bumps I hit on the way home. That was a couple days before Christmas. The night before, I was in my motel room looking at the corrugated blocks in the ceiling. I was counting the holes in the ceiling, while the other guys went to the show where they were playing Christmas songs. I thought, "I am married and I have a child, what am I doing here?" So I said to myself, "I am going home." The next morning, I went to the gas station where I kept my truck. I got it out of the garage and I started it up. I backed it up, loaded my car into the box, and came home. That was the end of my time on the Seaway.

Conclusion

THE CONSTRUCTION of the St. Lawrence Seaway was the greatest construction show ever performed. Twenty-two thousand men and their families migrated to Massena and Cornwall to spend four years constructing the waterway and power dam. During the project contractors and their employees tamed the Long Sault Rapids, excavated tenacious material, and dealt with extreme weather conditions. Working with sketchy plans and faulty concrete, these carpenters, engineers, laborers, and equipment operators completed a project that at the beginning of the twentieth century American and Canadian politicians and engineers had deemed expensive, impossible, and therefore unnecessary. The St. Lawrence Seaway and power dam construction exhibited the ability of men from different backgrounds and the leaders of American and Canadian construction companies and agencies to amicably complete a navigable waterway from Montreal and Lake Erie, full of numerous dredging, lock, and dam projects. Ontario Hydro and the Seaway Authority property agents and carpenters undertook the controversial relocation of thousands of Canadians along with their towns and homes. The project not only changed workers' and area residents' lives, but improved the navigational and power production facilities of both nations. It seems only appropriate to remember the men who worked so diligently to make the dream of America's Fourth Coast a reality. Without them, the St. Lawrence Seaway and Power Project would still be merely a drawing gathering dust on an aging engineer's desk.

The dedication of the St. Lawrence Seaway and Power Project marked the end of a century-long struggle to construct the largest power and navigational project in the world. President Dwight D. Eisenhower and Queen

Elizabeth II presided over the opening ceremony in Montreal on June 26, 1959, attended by dignitaries and members of the media from forty-eight countries. Following the brief festivities, the queen and the president boarded the royal yacht Britannia and proceeded down the river to the St. Lambert Locks and eventually on to Massena. Vice President Richard Nixon joined them the next day to watch the explosion that breached the remaining cofferdam, to dedicate the peace monument at the power dam in Massena, and to attend a luncheon at the Cornwallis Hotel in Cornwall. In his speech Eisenhower affirmed the project as a bilateral anti-Communist effort: "It is above all a magnificent symbol to the entire world of the achievement possible by two democratic nations peacefully working together for the common good."[1]

Downstream, before a crowd of fifty thousand, thirty tons of explosives blew two holes in the steel-celled A-2 dam that had held back the raging current of the St. Lawrence River for the last five years. Thirty minutes before the blast, police officers directed traffic near the project overlooks to prevent jams and accidents among the several thousand interested onlookers. Law enforcement also evacuated residents from homes

34. Vice President Nixon, Governor Nelson Rockefeller, Lewis G. Castle (administrator of the St. Lawrence Seaway Development Corporation), and others at the Eisenhower Lock. Courtesy of St. Lawrence Seaway Development Corporation.

35. Prince Philip (second from left), Mrs. Richard Nixon, Queen Elizabeth II, Vice President Richard Nixon, and others on the speakers' stand at the Eisenhower Lock. Courtesy of St. Lawrence Seaway Development Corporation.

within a two-mile radius of the demolition site. Once workers detonated the explosives, it took two and a half days to flood 38,000 acres between the Iroquois Control Dam and the main power structure to initiate power production.

The St. Lawrence Seaway officially opened for navigation on June 27, 1959. Vessels up to 730 feet long and 75.6 feet wide could transport iron ore, steel, and grain from the interior ports of both nations to waiting international cargo ships in Montreal. The final cost for the entire project, shared by the United States and Canada, was $1.2 billion dollars. The St. Lawrence Seaway and power dam replaced the old fourteen-foot-deep, thirty-lock canal system with 265 miles of twenty-seven-foot-deep channels, fifteen locks, and an international hydrodam the size of seven American football fields.

The dignitaries' luncheon—attended by 350 guests including New York Governor Averill Harriman; the Honorable Leslie Frost, prime

36. Vice President Richard Nixon on the speakers' stand at Eisenhower Lock. Courtesy of St. Lawrence Seaway Development Corporation.

minister of Ontario; Robert Moses, chairman of PASNY; and James Duncan, chairman of Ontario Hydro—marked the culmination of the four-day dedication festivities.

Dolores Kormanyos

When I was twenty-five, Edith, the head of the waitstaff at the Cornwallis, asked me to work at the dignitaries' banquet. One of my most cherished possessions is the menu from that meal that all of the attendees and the waitstaff took home. I had worked for Edith before, who ran the restaurant, and she liked the way I served. She wanted all of her most experienced staff to wait on the queen. She had called one of my girlfriends and asked if she wanted to work at the event as well.

The queen came down the Seaway in the *Britannia*. All of us who were working at the luncheon went down to watch her ship come in. The queen got off her yacht and drove to the power house. The staff of Ontario Hydro and PASNY had a special chair for the queen and President Eisenhower to sit in. After they signed the book with all of the names of the men who worked on the project and each gave a brief speech, Eisenhower left for another appearance and the queen was driven to Cornwall. The beginning of the luncheon was delayed for two hours because the fog had slowed down the royal yacht and also because the queen, who was pregnant with Edward, was not feeling well. Lionel Chevrier and his wife and another couple arrived at the scheduled time for the luncheon and sat in the dining room waiting for the queen. Even when she showed up, she went upstairs to her hotel room and freshened up. She also had a cup of tea sent up.

At the luncheon the Cornwallis chefs prepared nothing but the best food, which we served on the best china. I remember we used heavy crystal glasses and bowls and kept everything on ice in the basement so it didn't spoil. The guests had fruit cup supreme, which I carried on trays upstairs and in the air to each table. When I wasn't serving a dish or clearing away the dirty plates, I had to stand at attention. I had a difficult time focusing and keeping my composure with all of those famous people passing just four feet in front of me.

All of the cooks, the waitstaff, and the managers had to be screened by the mounted police to make sure we did not have criminal records. After I agreed to come and waitress I had to send the mounted police my name and date of birth. The 350 guests included all of the important leaders of the Western world at that time and someone could have had something against any one of them. Sitting at the head table was James Duncan, chairman of Ontario Hydro; Robert Moses; Leslie Frost, prime minister of Canada; Richard Nixon, vice president of the United States; Queen Elizabeth II; Richard Wigglesworth, United States ambassador to Canada; Nelson Rockefeller, governor of New York; John George Diefenbaker, Canadian minister of transport; Dr. J. A. Philips, mayor of Massena; and Rosario Brodeur, bishop of Alexandria; and all their wives.

Security was very tight both inside and outside the restaurant. All of these people stood on the sidewalk who wanted to see the queen. The town had erected big barricades that three mounted policemen guarded. Several others patrolled on foot including five stationed near the queen's car and two who never left her side.

Roy Simonds

A lot of high-ranking people came to visit the job during the construction phase. However, I will never forget the day of the dam dedication and the flooding of the power pool, which all of the key politicians from Canada and the U.S., including [the] queen, attended. I went down with my whole family and most of my neighbors to watch the beautiful royal yacht go through the Eisenhower Lock. Workers placed a red carpet leading to the specially constructed platform for all the important dignitaries to stand on at the dedication of the Robert Moses–Robert H. Saunders Power Dam. It was amazing to see something I had built in action.

Shirley Davis

My most memorable experience was the day they opened the Seaway. It was extremely exciting when they breached the cofferdam and flooded the waterway. Since my husband and I were living in Waddington, New York, we first stopped to see the explosion and then traveled toward Massena to catch a glimpse of the queen. It was so hot and foggy that day, we thought her yacht might never make it. It was yet another day of extreme weather and surprises on the project.

Robert Hampton

My neighbors and I all went to see the queen preside over the dedication of the two power plants that were side by side. My wife and I got up early and got the kids fed and dressed. We looked out the window and noticed that it was very foggy. We thought it might burn off with the daytime heat, but we figured that the queen might have to get off the yacht earlier then expected and travel the rest of the way by car. As I knew the project so well, I guessed where her docking point might be and I got it right. When we arrived, the queen got off right in front of where we were standing. She waved at us and then got into a waiting car.

Barbara Hampton

All of my friends pretended that they didn't want to go to the lock to see the queen. We had all decided that we had better things to do that day. Then all of a sudden everyone got queen fever. We all bought new dresses and wore our best hats. As the *Britannia* came up in the lock the queen's band was in the upper deck and they played "With a Little Bit of Luck." It seemed very appropriate due to the foul weather. When the queen got to the microphone, she said she was used to fog. She was so pretty and young at the time. It was amazing to see her and it turned out to be a gorgeous day. It was thrilling to see everything that our husbands had worked on finished and being praised by so many important people.

There was a photographer named Al Mellett who took pictures of the project for the Power Authority. He went on to be a part-time photographer for *National Geographic*. He was a wonderful man. He helped my husband, Bob, and me pick out a new camera and showed us how to use it. After the dedication ceremony ended, Al stopped by my house and told my friends and me about all of the politicians he had photographed. He also told us that the queen put lipstick on at the table after she ate her dinner. Finally, he described what clothes she had worn. We didn't have the heart to tell him that we had actually all seen her in person. He was so thrilled with the events of that day. It was quite an adventure for all of us.

Alfred Mellett
I got pictures of the dedication ceremony. I think my best one was of the queen and Eisenhower standing at the very center of the dam, which is the international dividing line and where the dedication ceremony took place. The Power Authority built a big overhanging structure jutting out from the dam to hold photographers and reporters so they had a view from the air of not only the main dam, but many of the structures downstream. I got a lot of photographs of the VIPs. All the big shots were there.

Janet Brodie
When the Seaway was being constructed, I attended Glengarry District High School in Alexandria. Once a year the students were taken in a bus to see the construction and the progress. At that time I was a teenager and the bus ride to Cornwall and the boys on the bus were more interesting than the bulldozers and the piles of dirt. When the Seaway was finally completed, I was married to Ed and living in Cornwall. Our daughter was two months old. The day of the flooding we were up bright and early. We

37. In 1958 the main power dam neared completion. The forebay was flooded; the power houses are both visible. The gantry cranes and earthmoving equipment remained. The area behind the dam will eventually be under water. Courtesy of Alfred Mellett and PASNY.

brought a blanket to sit on, a diaper bag, and proceeded to the site where we waited for three hours. Finally there was a puff of smoke and more waiting. Eventually a dribble of water appeared. We had been expecting something like a tidal wave and what a disappointment it was. We still laugh about it now.

John Dumas

I stood on the top of the new power dam with my mother on Inundation Day. Next to us was a fellow by the name of Robert Moses. When they blew the cofferdam holding the water back, there was a plume of smoke that went fifty feet in the air, but nothing else happened. I looked at Moses and said, "Is that all there is?" Mom could have shot me. I had a tremendous opportunity to have a front row seat for all of the important events.

Bill Goodrich

I attended the official opening with my friend Keith Henry, the chief hydraulic engineer for Ontario Hydro. The queen was standing ten feet away from us. Keith was kind of a rough and tough individual and his language was a little obscene. He started swearing under his breath and I asked him what was wrong after the blast was set off. He said, "Do you realize that the way that water is flowing down here it is going to take about ten days for it to get up to the dike where the queen is standing!"

Keith Henry

It was an extremely interesting job and I was very young. My fondest memory of the project was getting ready for the opening of the Iroquois Lock and the flooding of Lake St. Lawrence. On May 22, the first canaler *(The Calgarian)* bypassed the old Gallop Canal right down to the Iroquois Dam and was sent through the new lock. This was in effect the opening of the upper one-third of the new Seaway, and although there was not much general publicity about it, those of us who had been working on the project for years counted it as a very significant occasion.

On June 27, a big contingent of the fellows from the hydraulic section of Ontario Hydro arrived in Cornwall and we set up a roster to keep track of every facet of the forebay raising. Very important in this work was look-ing after the removal of the existing gauges and their replacement in their new permanent locations. We also needed extra help to follow the rise of the water at a number of the temporary gauges. Communication lines had to be set up so everyone could keep us informed at the power house control center and it was important that they knew when to move so they wouldn't get trapped by the rising water. I was provided with a helicopter for my own use on June 30. Later that afternoon, Gordon Mitchell told me to make a last patrol of the whole reservoir area to be sure that there were no people wandering about in the forebay and to see as well as I could that everything was ready.

It was a very impressive sight, the whole forty-mile stretch of river valley completely devoid of trees and no people or vehicles anywhere within the perimeter of the area to be flooded. The only exceptions were those workers at the old lock sites who were dismantling the equipment such as electric motors that were to be salvaged and removing and burning the timber lock gates and anything else that might form hazardous floating debris. When I got back, I was able to tell Mr. Mitchell that everything seemed to be in good order and ready for the next morning. The actual flooding was planned to start with the blasting and removal of Cofferdam A, which protected the power house. It was set for precisely 0600h on July 1. Mr. Mitchell said, "Keith, tomorrow morning, you make a tour of the area around the cofferdam, over to Long Sault Dam and up as far Aultsville and the head of the canal at Dickinson's Landing. We don't want to blow up or kill any tourists to start the proceedings, so go out at 0500h."

There were a lot of last-minute things to do and I didn't get away from the River Control Center all that night. At 5:00 A.M. I took off in the helicopter in very doubtful weather with thunderstorms all around and hail. Sure enough, we found a tent pitched less than half a mile from Cofferdam A and when we buzzed it a couple came out running, followed by four kids, all waving at us. I radioed this information to the control center and before we left we saw three police cars and a truck bouncing over the rough construction tracks to grab them and haul them out of harm's way. When I got back I made a quick check with all our people and then I told Mr. Mitchell that as far as the River Control Center was concerned, we were ready to go. He called all the construction directors, including those on the American side, and at six o'clock we all raced up to the head works to watch the big blast.

The blast was pretty awesome, but since it was three miles away, the appearance was not what we had expected. The blast was designed to lift and scatter the top two-thirds of the cofferdam. Once the blast was fired, water immediately began pouring over the remainder of the cofferdam and headed toward the power house. It was not exactly the awesome flood which we had all envisioned. Nevertheless, the forebay filling operation was under way. The closure of the tunnel ports in the northern section of the Long Sault Dam began to rise. The complete process took about three days, so the water rose about a foot per hour. From my point of view, the operation was a complete success. I'm sure there were plenty of glitches and many things we hadn't thought of, but there was certainly not a single major omission in our planning that gave rise to significant complaints. It all went as smooth as clockwork, and on the Fourth of July the navigation of the St. Lawrence River reopened through the new set of locks. In the end, the big adventure was over. The St. Lawrence Seaway and power system has run smoothly for almost fifty years.

Robert Hampton

My most unforgettable experience happened the day they completed all the work on the power dam. Workers opened the Iroquois Control Dam so the water would rise and fill the power pool. There was a drop of about eighty feet between Ogdensburg and the plant to generate power. In the upper end of the river there were a lot of farms and woods that were being flooded. Power Authority and Ontario Hydro officials had set up watches along the area when they opened the gates to make sure that none of the logs came up and blocked the way. I was in a raft with one of my co-workers near Iroquois and there were all these people along the highway watching the water rise. All of a sudden our motor died. I tried to put down an anchor, but we were in such deep water it didn't hit the bottom. The raft started to drift, so we tried to wave the American flag to signal the people on shore that we were in trouble. The people on the shore could see us, but they just kept waving back. Eventually a patrol boat arrived from the Canadian equivalent of the Coast Guard and brought us back to shore.

The blast that filled the power pool and the bypassing navigational facilities opened a new route for ships. However, at the time of its completion, economists and engineers characterized the Seaway locks and the channel depths as archaic and gave the waterway a lifespan of fifty years. A June 28, 1955, article in the Daily Standard-Freeholder *reported that 75 percent of the world's vessels could travel the Seaway. In actuality, only 4 percent of freighter captains could sail from the Great Lakes to Montreal fully loaded. In a 1995* Invention and Technology

38. Iroquois Lock in June 1958 was now taking shipping traffic around the completed Iroquois Dam. Courtesy of Alfred Mellett and PASNY.

article, Daniel McConville asserted that the Seaway's shallow channels and seven short and narrow locks made it obsolete even in the 1950s because the waterway's facilities could not accommodate most of the world's merchant shipping fleet. In its first season of operation, 5,289 vessels transited the new locks from July 4 to December 16, led by 4,696 commercial ships and 590 U.S. and Canadian government vessels. The advent of containers in the mid-1960s allowed larger amounts of cargo to be loaded on continually expanding oceangoing freighters and further limited the number of ships that could use the Seaway. These boats cannot enter the locks and their pilots cannot navigate the Seaway without getting stuck or running aground in some of the shallow shoals. Today the waterway can accommodate only 4 percent of the world's ships. In 2006, 3,742 ships traversed the waterway carrying 38,306 tons of cargo including grain, iron ore, coal, and bulk freight. However, the future of the waterway remains uncertain.[2]

Several ideas to enlarge the facility and extend the six-month shipping season have been proposed by the Corps of Engineers, politicians, and private engineering firms, including rebuilding the entire facility with deeper channels, widening the locks, and using innovative ice-breaking techniques. However, none of these ideas is presently economically or politically feasible. In 1998 the New York State Power Authority began a fifteen-year, $254-million life expansion and modernization program to replace or renovate most of the original equipment on the American side of the power dam. In 2005 the new Hawkins Point Visitors Center opened across the river from the power production facility. It replaced the observation deck at the Robert Moses Power Dam that had been closed since September 11, 2001, because of security concerns. The Corps and the St. Lawrence Seaway Development Corporation have launched a concrete refacing program on the Eisenhower and Snell Locks. The improvements to the security system depend on the available capital in both the United States and Canada to fund the expansion and also on the continued need for the waterway. The cost may outweigh the benefits as manufacturing in the Great Lakes region continues to decline.

39. The Iroquois Dam was fully operation in August 1958. Many see this structure as a safeguarding measure in case the level of Lake Ontario rises. Most of the time, the gates remain open. Courtesy of Alfred Mellett and PASNY.

Dolores Kormanyos

In the beginning engineers planned for there to be twin locks, one on the American side, one on the Canadian. That would have allowed more ships to use the waterway at the same time, but would not have taken care of the size issue. It was also suggested that the Eisenhower and Snell Locks should be longer and wider as ships were getting larger even back then. Initially the locks were always busy and ships were always on the river. Now the ships have gotten bigger than Seaway planners could have ever imagined. The locks are almost always empty and only occasionally do we see a ship pass by.

Garry Moore

I spent the last five years of my career with the Seaway Authority planning for future expansion. The dam business of building the Seaway began in 1927. In 1895 the Canadians established the requirements for the Welland Canal locks completed in 1932. The dimensions of that structure locked in the size of all others on the waterway because it didn't make any sense to make anything bigger than what already existed because it meant redoing the Welland section. One of the major problems with the lock dimensions was that everybody thought of the Seaway as an inland waterway and no one thought of it as a route that would be utilized by oceangoing vessels. The lakers are shorter and thinner than ocean vessels almost three to one. Widening the locks would have increased the longevity and traffic on the entire system and been a tremendous benefit.

Ambrose Andre

The existing project does not pay its own way. The Corps of Engineers is talking about twinning or enlarging the locks, which would cost a lot of money and not greatly increase the traffic on the waterway. It would allow bigger ships to go through the system, but you still need cargo to fill them. All these improvements would mean that the same amount of cargo would just be carried on fewer ships. The Corps was probably not too experienced in terms of designing waterways to accommodate ocean-going vessels because most of the Corps' work was inland.

The Seaway is not what it appears to be. It seemed like we (Americans) were forced into it in the first place. The Canadians said we are going to build the Seaway with or without you. As a gesture of goodwill if you would like to participate, you can. I think it was a great inconvenience for them to put up with us having half the Seaway. Probably the channels had to be on our side, but as far as the two locks go I think they would have been very happy to have both of the locks on their side. To them it was lifeblood and to us, the Midwestern people were the ones who wanted it for shipping their wheat. As far as the easterners, they wanted the freight business to remain with the railroads.

David Flewelling

The transportation end of it shuts down in the wintertime, so it isn't a year-round operation in terms of the Seaway part of it. It is only an eight-month-a-year operating season. The power aspect of it produces electricity 365 days a year. In a way, it is akin to the Erie Canal because it opened up the interior to marine transportation. The lock size and channel depth, however, inhibit the types of ships that can traverse the waterway. Pilots cannot maneuver big oil tankers through there because the width and the draft on the Seaway and the length of the locks are too small for today's container ships. It was a bit shortsighted for the Corps to build a bottleneck into a project when another twenty feet of concrete would not have killed them. It may be that they were looking down the pipe and saying that even if they made the locks ninety feet long, it still would have been too short for the larger ships that were being developed. The control was the Welland Canal. Really, anything that went through our locks had to go through the Welland, so there was no sense making ours any bigger because the Canadians had no plans for making any improvements. There are a lot of ships running aground all the while now. I don't know how they would get any of those big babies down through there.

James Romano

The Seaway had many problems right after it was completed. Since there is only one lock for ships going both ways, this can create long lines and delays. It also takes about ten minutes for lock operators to raise a ship. Ship captains have to wait until the ship exits the lock before another can enter. At the time of the construction, the Corps of Engineers wanted to build additional locks and make it two ways, but they didn't have the money. Now engineers are talking about adding new locks that would be wider and longer because some of the ships are now scraping the sides and damaging the concrete. If they made it bigger today, depending on the size of the locks, the traffic on the Seaway could increase.

The Seaway opened up the Great Lakes and brought ships to [the] heart of the country and that is what makes this country great. The men who designed the project and worked with and under me are unsung heroes. We worked in horrible conditions and never asked for nothing. Today, I don't think there are dedicated and skilled workers who would go

40. Two passenger ships, the *South American* and the *North American* were the first to be lowered by the Eisenhower Lock on September 13, 1958. Courtesy of Alfred Mellett and PASNY.

through what we did to upgrade that facility and that is a shame. The U.S. needs to do something to remain on top; I just don't know if the government or the environmentalists are willing to let that happen. The Seaway will probably crumble until it's useless before the politicians can agree on how to improve it.

Notes | Bibliography | Index

Notes

Introduction

1. M. W. Oettershagen, deputy administrator of the St. Lawrence Development Corporation, "St. Lawrence Seaway—Fact and Future," May 5, 1959, 2, St. Lawrence Seaway Collection, St. Lawrence University, Canton, New York (hereafter cited as SLUSC).

2. William R. Willoughby, *The St. Lawrence Waterway: A Study in Politics and Diplomacy* (Madison: Univ. of Wisconsin Press, 1961), 264.

3. "Historian First-Hand Observer of Project," *Massena Observer,* June 26, 1984, 10; copy of article in author's possession.

4. Lionel Chevrier, *The St. Lawrence Seaway* (Toronto: Macmillan of Canada, 1959), 5.

1. The Binational Political Debate

1. William T. Easterbrook and Hugh Aitken, *Canadian Economic History* (Toronto: Macmillan of Canada, 1958), 553.

2. Willoughby, *St. Lawrence Waterway,* 59.

3. N. R. Danielian, *The St. Lawrence Survey, Part I, History of the St. Lawrence Project* (Washington, D.C.: U.S. Department of Commerce, 1939), 2–3; Alexander Wiley, *St. Lawrence Seaway Manual: A Compilation of Documents on the Great Lakes Seaway Project and Correlated Power Development* (Washington, D.C.: U.S. Government Printing Office, 1955), 176; and T. L. Hill, *The St. Lawrence Seaway* (New York: Frederick Praeger, 1959), 63.

4. George W. Stephens, *The St. Lawrence Seaway Waterway Project: The Story of the St. Lawrence River as an International Highway for Water-borne Commerce* (Montreal: Louis Carrier and Co., 1930), 156.

5. R[ichard] R[eeve] Baxter, ed., *Documents on the St. Lawrence Seaway* (New York: Frederick Praeger, 1961), 7.

6. G. P. de T. Glazebrook, *A History of Transportation in Canada* (Toronto: Ryerson Press, 1938), 432.

7. Press release from the St. Lawrence Seaway Development Corporation, Feb. 10, 1958, SLUSC.

8. Bertrand H. Snell, 1955 interview with former *Courier-Freeman* editor T. P. North, quoted in "Bertrand Snell is St. Lawrence Seaway's 'Forgotten Man,'" *Massena Observer*, July 10, 1984, 5, article in author's possession.

9. H.R. Bill 3778, 65th Cong., 1st sess., Apr. 24, 1917, supplied by SUNY Potsdam College Archives and Special Collections, C9, folder 2.

10. Conrad Payling Wright, *The St. Lawrence Deep Waterway: A Canadian Appraisal* (Toronto: Macmillan of Canada, 1935), 27–41.

11. Chevrier, *The St. Lawrence Seaway*, 27; and Willoughby, *St. Lawrence Waterway*, 91–94.

12. "The Great Lakes St. Lawrence Seaway Project: Digest of Presidential Papers, Messages and Official Statements, The Power Authority of the State of New York," no date, 24–25, SLUSC.

13. Stephens, *St. Lawrence Seaway Waterway Project*, 195; Willoughby, *St. Lawrence Waterway*, 97–98.

14. Willoughby, *St. Lawrence Waterway*, 167–68.

15. Stephens, *St. Lawrence Seaway Waterway Project*, 106–8; Hill, *St. Lawrence Seaway*, 64–65; and R. B. Willis, "The St. Lawrence Seaway: Economics for Canadians," *Quarterly Review of Commerce*, 1941, 255.

16. Stephens, *St. Lawrence Seaway Waterway Project*, 208.

17. Ibid., 217 and 219; Wright, *St. Lawrence Deep Waterway*, 65 and 99.

18. Willoughby, *St. Lawrence Waterway*, 111.

19. Ibid., 167–68; Stephens, *St. Lawrence Seaway Waterway Project*, 226–28.

20. Unpublished papers of Franklin D. Roosevelt, file 156, Roosevelt Memorial Library, Hyde Park, New York.

21. Willoughby, *St. Lawrence Waterway*, 105–6 and 214; Carleton Mabee, *The Seaway Story* (New York: Macmillan, 1961), 145.

22. Mabee, *Seaway Story*, 93–94.

23. Keith Fleming, *Power at Cost: Ontario Hydro and Rural Electrification, 1911–1958* (McGill-Queen's Univ. Press, 1992), 21.

24. Willoughby, *St. Lawrence Waterway*, 105–6 and 214; Mabee, *Seaway Story*, 145.

25. Willoughby, *St. Lawrence Waterway*, 145–46; Hill, *St. Lawrence Seaway*, 65–66.

26. "The Great Lakes St. Lawrence Seaway Project: Digest of Presidential Papers."

27. Willoughby, *St. Lawrence Waterway*, 145–46; Hill, *St. Lawrence Seaway*, 65–66.

28. White House Press Conference, Mar. 14, 1934, cited in Mabee, *Seaway Story*.

29. Hill, *St. Lawrence Seaway*, 66; Willoughby, *St. Lawrence Waterway*, 145–49.

30. "The Real Facts on the St. Lawrence Seaway. A Speech of Honorable Bertrand Snell of New York in the House of Representatives, Wednesday, January 17, 1934," *Congressional Record*, 73rd Cong., 2nd sess., 1.

31. "Massena Allies Other Groups in Power Plan," *Massena Observer*, Apr. 9, 1931, 1, 6.

32. "Massena Sends Men to Work For Seaway," *Massena Observer*, Jan. 18, 1934, 5.

33. "The Real Facts on the St. Lawrence Seaway," 8.

34. *Massena Observer,* Jan. 18, 1934, 1; *Watertown Daily Times,* Apr. 7, 1962, 3.

35. Willoughby, *St. Lawrence Waterway,* 161, 178, and 180.

36. Mabee, *Seaway Story,* 129.

37. Willis, 267–68.

38. "The St. Lawrence Seaway and Power Project for Defense and Commerce—Excerpts from Testimony before the Rivers and Harbors Committee, U.S. House of Representatives, July 1941," 8, SLUSC.

39. Ibid.

40. Ibid., 16.

41. Mabee, *Seaway Story,* 134–35.

42. Hill, *St. Lawrence Seaway,* 67.

43. *Massena Observer,* Sept. 5 and 7, 1967.

44. R. A. Newton, president of the Northern Federation of Chambers of Commerce, to the Honorable George D. Aiken, Apr. 26, 1945, with five-page attachment, St. Lawrence Univ. Library Archives, Seaway Collection.

45. *Watertown Daily Times,* Aug. 30, 1957.

46. Ibid.

47. Ontario Hydro, St. Lawrence Seaway Power Project brochure, Aug. 1956, 2, in author's possession.

48. *Cornwall Standard-Freeholder* (Cornwall, Ontario), May 28, 1958, 18.

49. Ibid., May 15, 1954; Leslie Roberts, "Canada and the St. Lawrence Seaway," *Canadian Letter,* Dec. 1951, 127–29.

50. Chevrier, *The St. Lawrence Seaway,* 41–43.

51. Ibid., 42–43.

52. Ibid., 47–48.

53. "Dr. Rollin A. Newton Urges Seaway Support in Letters to House Committee Members," *Massena Observer,* Apr. 30, 1951, 1.

54. *Massena Observer,* Apr. 12, 1951.

55. "Newton Gets Answer from Truman Letter," *Massena Observer,* Oct. 11, 1951, 1.

56. U.S. Senator Charles W. Tobey, address delivered before the National Rural Electric Cooperative Association, Cleveland, Ohio, Jan. 30, 1951, SLUSC; Thomas W. Nahl, "It's Time to Stop Fiddling," *Economic Outlook,* Jan. 30, 1948, and "The St. Lawrence Waterway Long Overdue," *Economic Outlook,* Feb. 1951.

57. Nov. 9, 1945, draft of article for *Democratic Digest,* 7, SLUSC.

58. "Statement of the Honorable Dean Acheson, Secretary of State, in support of H.J. Res. Approving the St. Lawrence Seaway and Power Project before the House Committee of Public Works, Tuesday, Feb. 20, 1951," SLUSC. These arguments are also included in a reprint of Senator George Aiken's speech to the U.S. Senate in *Great Lakes Outlook,* Jan. 1951, 8.

59. Wiley, *St. Lawrence Seaway Manual,* 67; memo to senators from Wiley, 1951, SLUSC.

60. Willoughby, *St. Lawrence Waterway,* 228.

2. The Project

1. John Brior, *Taming of the Sault: A History of the St. Lawrence Power Development—Heart of the Seaway* (Watertown, N.Y.: Hungerford-Holbrook Co., 1960), 14.

2. Mabee, *Seaway Story,* 172.

3. *The International St. Lawrence Seaway and Power Development, Billion Dollar Story,* vol. 5 (Massena, N.Y.: St. Lawrence Valley Souvenir Co., 1962), 12–13; Jacques Lesstrang, *Seaway: The Untold Story of North America's Fourth Seacoast* (Vancouver, B.C.: Evergreen Press, 1976), 82–89.

4. *Billion Dollar Story,* 12–13; Lesstrang, *Seaway: The Untold Story,* 82–89.

5. *Billion Dollar Story,* 12–13; Lesstrang, *Seaway: The Untold Story,* 82–89.

6. Robert A. Caro, *The Power Broker* (New York: Vintage Books, 1974), 1–20.

7. Ibid.

8. Ibid, 580, 595–97, and 709.

9. Ibid.

10. "4,200 Foot Cofferdam Links Island," June 28, 1955, 20, copy of article in author's possession.

11. "Equipment Is Worth Millions," *Daily Standard–Freeholder,* July 10, 1956, 5.

12. Power Authority of the State of New York, *St. Lawrence Power Project Data and Statistics,* copy of article in author's possession.

13. *Billion Dollar Story,* 62; Power Authority of the State of New York, *St. Lawrence Power Project Data and Statistics.*

14. "Gentleman Tops Them All," *Daily Standard–Freeholder,* June 29, 1957, 37; and "Giant Gantry Cranes Play Vital Role," *Daily Standard–Freeholder,* July 10, 1956, 3.

15. "Draglines Float to Seaway," *Excavating Engineer,* Aug. 1955, 42; "Huge Dragline at Work on Seaway," *Watertown Daily Times,* June 19, 1956, 5.

16. "Draglines Float to Seaway," 42; "Huge Dragline at Work on Seaway," 5.

17. "Hartshorne Brothers Familiar Figures in Seaway Valley," *Daily Standard–Freeholder,* July 10, 1956, 20.

3. The Workforce

1. "Uhl, Hall and Rich Firm Making Engineering Studies, Designs for Power Authority," *Massena Observer,* Aug. 8, 1955, 1.

2. *Federal Irrigation Project, Bureau of Reclamation* (Washington, D.C.: U.S. Government Printing Office), 1–4.

3. "Hungry Horse Project, Montana," <www.usbr.gov/dataweb/html/hhorse.html>.

4. "U.S. Corps of Engineers Is Getting Bright Young Engineers with Ideas," *Massena Observer,* Aug. 1, 1955, 1.

5. "TVA: The Enduring Legacy," <www.tva.gov/heritage/fdr/index.html>.

6. Ibid.

7. Ibid.

8. "Director of Project Hydro Vet," *Daily Standard-Freeholder,* July 10, 1956, 32.

9. "Employment on the Power Project Soon to Rise to 10,000 Figure; 5,900 Worked on American Side of St. Lawrence River in July 1956," *Massena Observer,* Feb. 18, 1957.

10. *Billion Dollar Story,* 14–19.

11. Daniel McConville, "November Pour," *The Quarterly,* Apr. 1988, 3.

12. U.S. Court of Appeals, 2nd Circuit, no. 78, docket 25613, *Morrison-Knudsen Company, Inc., Walsh Construction Company and Perini-Quebec, Inc., Joint Venturers doing business as Robinson Bay Lock Constructors; Morrison-Knudsen Company, Inc., B. Perini and Sons, Inc., Walsh Construction Company and Utah Construction Company, a Joint Venture and Selby Drilling Corp., Petitioners, v. National Labor Relations Board,* argued Dec. 11, 1959, decided Mar. 2, 1960.

13. "River Projects Stalled as 1,200 Workers Strike," *Watertown Times,* Mar. 12, 1956, 22; "State Mediation Board Enters Project Strike," *Watertown Times,* Mar. 13, 1956, 19; "Project Activity Resumed as Workers Get Increase," *Watertown Times,* Mar. 19, 1956, 10.

14. U.S. Court of Appeals, 2nd Circuit, no. 78, docket 25613.

15. Ibid.

16. *St. Lawrence Power,* final ed., Sept. 5, 1958, 4, 12, and 21.

17. Ibid.

4. Life on the Job

1. Lesstrang, *Seaway: The Untold Story,* 69.

2. "Safety Record on Project 'Model of Achievement,'" *Watertown Daily Times,* July 9, 1957, 8.

3. *Courier and Freeman,* July 10, 1956, 4.

4. Sam Agati, Massena, New York, Mar. 8, 1989, interview by author.

5. Walter Gorrow, Massena, New York, Mar. 8, 1989.

6. "Price of St. Lawrence Seaway Can Be Measured in Men Too," *Chicago Daily News,* June 26, 1958, Seaway section, B6.

7. "New Safety Techniques Cut Accidents in Construction," *Chicago Daily News,* June 26, 1958, B7.

5. Construction Dilemmas

1. William Becker, *From the Atlantic to the Great Lakes: A History of the U.S. Army Corps of Engineers and the St. Lawrence Seaway* (Washington, D.C.: Historical Division, Office of the Chief of Engineers, U.S. Army Corps of Engineers, 1984), 44–51.

6. Off the Project

1. Douglas T. Miller and Marion Nowak, *The Way We Really Were* (New York: Doubleday, 1977), 8.

2. Robert H. Young and Nancy K. Young, *The 1950s* (Westport, Conn.: Greenwood Press, 2004), 21.

3. "Getting Acquainted Easy, Engineer Declares in Speech Given at Toastmasters Club," *Massena Observer*, Oct. 22, 1956.

4. Alan Emory, "Project Brings Big Change to Massena," *Watertown Daily Times*, June 6, 1956.

5. Dick Peer, "Peering at Massena," *Massena Observer*, Nov. 7, 1955.

6. Young and Young, *The 1950s*, 126 and 130.

7. David Halberstam, *The Fifties* (New York: Fawcett-Columbine, 1993), 131.

8. Ibid., 132 and 145.

9. Young and Young, *The 1950s*, 109.

Conclusion

1. "Lock Dedicated as Monument to Peaceful World," *Watertown Daily Times*, June 6, 1956, 1.

2. *The St. Lawrence Seaway Traffic Report Historical Tables, 1959–1992*, 4–5, <www.grand-slacs-voiemaritime.com/en/pdf/traffic_report_hist.pdf>; Saint Lawrence Seaway Development Corporation, *Seaway System—The Seaway—Facts and Figures—Traffic*, Traffic Reports, 2005, 2006, 2007, <www.greatlakes-seaway.com/en/seaway/facts/traffic/index.html>; and *The St. Lawrence Seaway Management Corporation Annual Report, 2005/2006*, SLUSC.

Bibliography

Newspapers

Chicago Daily News
Christian Science Monitor
Cornwall Standard-Freeholder
The Courier and Freeman
Massena Observer
Watertown Daily Times

Primary and Secondary Sources

Baxter, R[ichard] R[eeve], ed. *Documents on the St. Lawrence Seaway.* New York: Frederick Praeger, 1961.

Becker, William. *From the Atlantic to the Great Lakes: A History of the U.S. Army Corps of Engineers and the St. Lawrence Seaway.* Washington, D.C.: Historical Division, Office of the Chief of Engineers, U.S. Army Corps of Engineers, 1984.

Brior, John. *Taming of the Sault: A History of the St. Lawrence Power Development— Heart of the Seaway.* Watertown, N.Y.: Hungerford-Holbrook Co., 1960.

Caro, Robert A. *The Power Broker.* New York: Vintage Books, 1974.

Chevrier, Lionel. *The St. Lawrence Seaway.* Toronto: Macmillan of Canada, 1959.

Cullen, Donald E. "Union Wage in Heavy Construction: The St. Lawrence Seaway." *American Economic Review* 49 (Mar. 1959): 68–84.

Danielian, N. R. *The St. Lawrence Survey, Part I, History of the St. Lawrence Project.* Washington, D.C.: U.S. Department of Commerce, 1941.

Dempsey, William. *Advantages and Necessity of an All-American Waterway from the Great Lakes to the Atlantic.* Speech presented to the House of Representatives, Mar. 1926. Washington, D.C.: U.S. Government Printing Office, 1926.

Draft of article for *Democratic Digest*, Nov. 9, 1945, 7, SLUSC.

Easterbrook, W. T., and Hugh Aitken. *Canadian Economic History*. Toronto: Macmillan of Canada, 1958.

Federal Irrigation Project, Bureau of Reclamation. Washington, D.C.: U.S. Government Printing Office, 1950.

Fleming, Keith. *Power at Cost: Ontario Hydro and Rural Electrification, 1911–1958*. Toronto: McGill-Queen's Univ. Press, 1992.

Glazebrook, G. P. de T. *A History of Transportation in Canada*. Toronto: Ryerson Press, 1938.

Halberstam, David. *The Fifties*. New York: Fawcett-Columbine, 1993.

Hill, T[heodore] L[ewis]. *The St. Lawrence Seaway*. New York: Frederick Praeger, 1959.

H.R. Bill 3778, 65th Cong., 1st sess. Supplied by SUNY Potsdam College Archives and Special Collections, C9, folder 2.

"Hungry Horse Project, Montana."<www.usbr.gov/dataweb/html/hhorse.html>.

The International St. Lawrence Seaway and Power Development, Billion Dollar Story. Vol. 5. Massena, N.Y.: St. Lawrence Valley Souvenir Company, 1962.

Keesbury, Forrest. "The Role of Dwight D. Eisenhower in the Development of the St. Lawrence Seaway." M.A. thesis, Bowling Green State Univ., 1965.

Lesstrang, Jacques. *Seaway: The Untold Story of North America's Fourth Seacoast*. Vancouver, B.C.: Evergreen Press, 1976.

Mabee, Carleton. *The Seaway Story*. New York: Macmillan, 1961.

May, Elaine Tyler. *Homeward Bound: American Families in the Cold War*. New York: Basic Books, 1988.

McConville, Daniel. "November Pour." *The Quarterly*, Apr. 1988.

Memo to senators from Wiley, dated 1951. St. Lawrence Seaway Collection, St. Lawrence University, Canton, New York (hereafter cited as SLUSC).

Miller, Douglas T., and Marion Nowak. *The Way We Really Were*. New York: Doubleday, 1977.

Newton, R. A., president of the Northern Federation of Chambers of Commerce, to the Honorable George D. Aiken, Apr. 26, 1945, with five-page attachment, SLUSC.

O'Dwyer, William. Speech to the New York State Commerce and Industry Association, Jan. 29, 1948, SLUSC.

Oettershagen, M. W. "St. Lawrence Seaway—Fact and Future," May 5, 1959, SLUSC.

Ontario Hydro. St. Lawrence Seaway Power Project brochure, Aug. 1956, copy in author's possession.

————. *St. Lawrence Power.* Final ed. Sept. 5, 1958, copy in author's possession.

"The Real Facts on the St. Lawrence Seaway. A Speech of Honorable Bertrand Snell of New York in the House of Representatives, Wednesday, January 17, 1934." *Congressional Record,* Seventy-Third Cong., 2nd sess.

Roberts, Leslie. "Canada and the St. Lawrence Seaway." *Canadian Letter,* Dec. 1951.

Roosevelt, Franklin D. Unpublished papers, file 156, Roosevelt Memorial Library, Hyde Park, New York.

St. Lawrence Seaway Development Corporation. "Power Authority of the State of New York St. Lawrence Power Project Data and Statistics." Press release, Feb. 10, 1958, SLUSC.

"Statement of the Honorable Dean Acheson, Secretary of State, in support of H.J. Res. Approving the St. Lawrence Seaway and Power Project before the House Committee of Public Works." Tuesday, Feb. 20, 1951, SLUSC.

Stephens, George W. *The St. Lawrence Seaway Waterway Project: The Story of the St. Lawrence River as an International Highway for Water-borne Commerce.* Montreal: Louis Carrier and Co., 1930.

The St. Lawrence Seaway and Power Project for Defense and Commerce—Excerpts from Testimony before the Rivers and Harbors Committee, U.S. House of Representatives, July 1941, SLUSC.

"TVA: The Enduring Legacy." <www.tva.gov/heritage/fdr/index.html>.

Wiley, Alexander. *St. Lawrence Seaway Manual: A Compilation of Documents on the Great Lakes Seaway Project and Correlated Power Development.* Washington, D.C.: U.S. Government Printing Office, 1955.

Willis, R. B. "The St. Lawrence Seaway: Economics for Canadians." *Quarterly Review of Commerce,* 1941.

Willoughby, William. *The St. Lawrence Waterway: A Study in Politics and Diplomacy.* Madison: Univ. of Wisconsin Press, 1961.

Wright, Conrad Payling. *The St. Lawrence Deep Waterway: A Canadian Appraisal.* Toronto: Macmillan of Canada, 1935.

Young, Robert H., and Nancy K. Young. *The 1950s.* Westport, Conn.: Greenwood Press, 2004.

Index

Italic page numbers denote illustrations.